张振山 著
[美] 欧思博 译

画谈潜园

中国园林在德国

By Zhang Zhenshan
Evan Osborne Tr.

A Chinese Garden in Germany

访谈 Interview

江岱（Jiang Dai）× 张振山（Zhang Zhenshan）

江岱（Jiang Dai）

同济大学出版社编审、副总编辑

Senior Editor and Deputy Chief Editor, Tongji University Press

张振山（Zhang Zhenshan）

原任同济大学建筑系教授

原任上海同济城市规划设计研究院总建筑师

国家一级注册建筑师

Professor Emeritus of Architecture, Tongji University and Former Chief Architect, Shanghai Tongji Urban Planning and Design Institute

Registered Architect

访谈时间 / 地点：2019 年 9 月，上海市松江区

Interview conducted in September 2019, in Songjiang, Shanghai

江： 著名建筑师黑川纪章曾讲过，建筑是本历史书，在城市中漫步，应该能够阅读它，阅读它的历史、它的意韵。在建筑设计中既要有实用性，又要融入人文性，这对建筑师本身的文化素养要求就非常高。园林意境也是设计师文化素养的流露，我们也能从您的设计作品里看出来。没有深厚的文化底蕴，造不出园子迷人的意境以及深刻的内涵？

Jiang: The famous architect Kisho Kurokawa has said that architecture is a history book. When strolling through a city, one should be able to "read" it, and through it to understand the city's history and charm. Architectural design needs to be both practical and humanistic. And this requires in the architect a very high degree of cultural sophistication. A garden's artistic conception is also a reflection of its designer's cultural sophistication, which we can see in your works. Lacking such a deep cultural foundation, is it possible to create an enchanting mood, a profound essence?

张： 我不懂文学，但我在天津大学读书的时候，王学仲老师（1925—2013，著名书画家）给我们讲了一两年的中国美术史、世界美术史。他讲了上百个故事，对我非常受用，也是自此对中华文化有了深入的了解。老师既是"黾学学派"的创始人，又是"黾学"最好的践行者。他提出学术主张，形成包含有哲学、美学、书学、文学、诗词学和绘画学的完整的"黾学"学术体系。他的山水艺术气质来源于他熟谙汉文化后流露出的那份从容和定力，也在于他东西技法融而合之的均衡柔和。十几年前，我从三门峡到函谷关考察，函谷关也是老子著述道家学派开山巨著《道德经》的灵谷圣地。里面有很多石碑，突然我看到有个石碑是王学仲老师题字的，我看了非常激动。后来，我又去了杭州于谦祠，又看到王学仲老师题写的匾额。深厚的文化底蕴对设计师来说是非常重要的，作为后辈，能有这么好的老师真的非常自豪。

建筑和文化有深厚的渊源，教学里面怎么去发掘这个东西，非常重要。其实大家都在忙着搞科研、搞项目，真正坐下来好好研究学问的时间真的很少。现在的年轻人对新事物接受快，很容易受到西方文化的影响，而忽视了对中华文化的学习和研究。很多中国传统文化的优秀内容，都渐渐失散了，造成"中国文化本位"的缺失。

Zhang: I do not know literature, but when studying at Tianjin University, Prof. Wang Xuezhong (famous calligrapher and painter, 1925–2013) taught us one or two years of Chinese art history and world art history. He told us more than 100 stories. I found them very useful, and this gave me a much deeper understanding of Chinese culture. The professor was the originator and the foremost practitioner of the Mianxue school. He put forward his own academic ideas and created a comprehensive school of thought that included philosophy, aesthetics, calligraphy, literature, poetics, and painting. His artistic temperament stemmed from the calmness and definitiveness that emerged after he enriched Han culture, and from his well-balanced blending of Eastern and

五

1

2

3

4

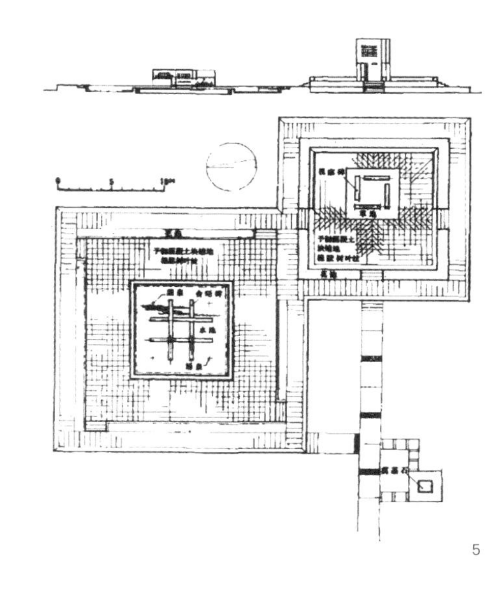

5

1. 潜园竣工留影
2. 在函谷关与王学仲老师题字的石碑合影
3. 与鲁尔大学领导商议潜园工程
4. 上海市人民英雄纪念塔(主设计师之一)
5. 周总理视察西双版纳、发展中缅友谊纪念碑设计图

Western techniques. More than a decade ago, I went from Sanmenxia [a city in Henan] to Hanguguan [a famous mountain pass in where is today Henan], Hanguguan being the spiritual location where Lao Tzu wrote the Taoist school's masterpiece, the *Tao Te Ching*. There were many steles inside, and at one point I suddenly saw a stone monument with the inscription written by Professor Wang; I was thrilled to see it. Later in my life, I went to Yu Qian's Memorial Temple in Hangzhou and again saw this plaque. This kind of profound cultural heritage is very important for architects. As a member of the generation that followed that of Prof. Wang, I am really proud of having such a good teacher.

Architecture and culture have a common, profound source, and how to bring this out in teaching is profoundly important. But these days everyone is busy doing research projects, and there is very little time to sit down and look into such questions. Now young people are quick to accept new things and Western views, and neglect the study of and research into Chinese culture. A lot of outstanding traditional Chinese culture has been lost, and this has unfortunately resulted in the loss of a "Chinese cultural standard".

江：建筑是凝固的音乐，建筑是应该有灵魂的。文化的素养，决定了人的气质；文化的积淀，是否就是建筑的灵魂呢？陈从周曾说，中国园林是文人园，实基于“文”。这位并非科班出身的园林大家，将文学、书画与园林融会贯通，走出一条饱含中国人文情怀的园林之路。您前面也提到文化底蕴非常重要，您能给我们谈一谈建筑与文化的关系吗？

Jiang: Architecture is a kind of robust music, and it should have a soul. And cultural accomplishment determines one's temperament. Is the accumulation of culture the soul of architecture? Chen Congzhou once said that Chinese gardens are gardens of literati, and are actually based on a sort of "literature". The landscape architect, who actually didn't major in architecture as a student, still integrates literature, calligraphy and painting into the gardens he designs, and leaves behind a number of gardens full of Chinese humanity. You also mentioned that cultural heritage is very important. Can you tell us about the relationship between architecture and culture?

张：“求木之长者，必固其根本”，深挖建筑，下面是文化与历史，是语言与精神。雄伟的建筑在于其精神脊梁，历久的建筑在于其无声语言。建筑要有人文之根，要具备自己的文化主体意识。在这个基础上的兼容并蓄才有活力，在这个基础上的传承与创新才有意义。

冯纪忠先生设计的上海松江方塔园中的何陋轩，借用唐代大诗人刘禹锡的《陋室铭》命此轩为“何陋轩”，极富哲理，很有文化意蕴，如果叫“方塔茶室”，这就完全不一样了。陈从周先生不仅对于古建筑、古园林理论有着深入的研究、独到的见解，还参与了大量实际工程的设计建造，同时他还是一位知名的散文作家和画家，是张大千先生的入室弟子。正是因为他所具有的多重身份和深厚广博的文化积淀，才能够引领读者在古典建筑和文化的长廊里徜徉。眼界高时无物碍，陈从周先生提议仿造苏州网师园中的殿春簃，在美国纽约大都会博物馆内兴建一座中国庭园（阿斯特园，又称“明轩”），其想法与眼界令人击节。将中国园林艺术推向世界，他是现代第一人。文化之于建筑是非常重要的，在当前的专业教学里却没有很好地融入，是我认为比较遗憾的一点。

Zhang: It has been said that “If one wants the tree to grow fully, one must tend to its roots”. In other words, architecture must delve deeply into history, language and the spiritual. The essence of a majestic building is its spiritual backbone, and that of a long-standing building lies in its “unspoken language”. Architecture should be rooted in humanism, and have its own culturally conscious primary themes. This kind of compatibility is vital, and only inheritance and innovation with these roots are meaningful.

In the Square Pagoda Park in Songjiang, Shanghai, Mr. Feng Jizhong designed a building named Helou Pavilion, which was borrowed from the Tang Dynasty poet Liu Yuxi’s “My Humble Abode” (Loushi Ming) , a phrase replete with philosophical and cultural significance; had it instead been called something like the “Square Pagoda Tea House” it would have been completely different. Mr. Chen Congzhou not only engaged in deep study of ancient architecture and the theory of ancient gardens and thus derived unique insights, but also participated in a great deal of practical engineering design works, even while he was a well-known prose writer and painter; he was a disciple of Mr. Zhang Daqian. It was only because of Mr. Chen’s multiple roles and his extensive accumulation of culture that he could lead the reader in ambling through the classical architecture and the culture of this pavilion. And perceiving no obstacles to setting his sights high, Mr. Chen Congzhou proposed to imitate the Peony Study (Dianchunyi) in the Master of the Nets Garden in Suzhou via building another Chinese garden (the Astor Chinese Garden Court, also known as Ming Xuan) in the Metropolitan Museum of Art in New York. His ideas and vision strike us as quite admirable. He was the first person in the modern era to bring the art of the Chinese garden to the world. Culture is critical to architecture, but it is a pity that culture is not well integrated into current professional training.

江： 我们知道您做了很多设计，对您来说印象最深刻的有哪些作品，能和我们分享一下吗?

Jiang: We know you’ve designed many works. Can you share with us which ones have left the deepest impression on you?

张: 我做的设计有很多, 印象最深刻的有三个: 第一个是上海市人民英雄纪念塔，第二个是在西双版纳做的周总理视察西双版纳、发展中缅友谊纪念碑，第三个是在德国做的一个中国园林（潜园）。我对前面两个作品做一些解释：

在上海黄浦江与苏州河交汇处的黄浦公园内，矗立着上海市人民英雄纪念塔。三十多年过去了，设计纪念塔的情景仍时常会浮现在我的眼前，仿佛就在昨天。

上海解放初期，有关各界就酝酿着建立一座上海人民英雄纪念塔。由于选址多次变更，前两次的中标设计都未能实施。1987 年，上海市政府决定在黄浦公园内选址建造上海市人民英雄纪念塔，并开展了设计竞赛。当时上交的 100 个方案中，我们的设计一举夺魁，成为实施建造方案。纪念塔设计最大的难题，是在黄浦公园内沿黄浦江与苏州河的三角地段，如何容纳一座标志性的宏大主题建筑。若处理不好，就会成为公园内雕塑级的小品。当我们寻找立意构思时，想到它应该是上海的母亲河苏州河、黄浦江畔的一颗明珠。有了明珠的概念，我们的思绪顿时豁然开朗，形成了“圆岛”的设计。“圆岛”的构图不单在造型理念上与周围形成对比，同时也使纪念塔处于黄浦公园及外滩旅游带的“龙头”地位，从空中鸟瞰更能感受到其整体美的设计理念！

位于西双版纳自治州首府景洪，云南热带作物研究所三号橡胶林中的纪念碑，由纪念周总理视察西双版纳的“视察碑”和纪念中缅两国人民友好情谊的“友谊碑”组成，由我和司马铨老师作为主设计师参与完成。友谊碑的立意受两手相握形态启发，以四个相同的折尺形单元构件相互穿插，作“井”字形布局，翩翩起舞于水池碧波之上，体现了两国互相尊重、和平共处的意愿，以及中缅人民源远流长、世代相传的胞波情谊。与友谊碑呼应的是垂直挺立于高台之上的视察碑，碑体由四块竖立的曲尺形构件四方环立组成。竖向布局继承中国石碑营造的传统手法，构件造型象征着四棵橡胶大树，寓意橡胶事业如茁壮成长的阔叶大树，蓬勃发展。此外，在碑台引道两侧，设奠基石，志建碑经过，使总体气氛更为谐调。

设计本天成，妙手偶得之。有时建筑设计师常常等待灵感的降临，灵感一至，便付诸纸面。在我们那个年代，建筑设计图纸不是出自冷冰冰的电脑软件，而是出自活生生的人的手笔。这是有人文的温度的，这种温度把一张张稿纸熨帖舒展，使其能经得起岁月长河的冲蚀。

Zhang: I have done a lot of design works, and there are three projects that have left the deepest impression on me. The first is the Shanghai People's Heroes Memorial, the second is the monuments to Premier Zhou's inspection tour in Xishuangbanna and the friendship of China and Myanmar. The third project is the Chinese garden in Germany. Let me explain the first two works. In Shanghai's Huangpu Park, where the Huangpu River and the Suzhou River intersect, stands the Shanghai People's Heroes Memorial Tower. Thirty years later, the imagery from when I participated in the design of this memorial tower still seems like it's right before my eyes, as if it happened only yesterday. Just after the liberation of Shanghai, all sorts of people were considering how to build this tower. Because the chosen specifications kept changing, the first two plans selected could not be realized. In 1987 the Shanghai city government decided to select a tower for the city's Huangpu Park, and the design competition was then carried out. Of the 100 designs submitted, ours won and began to be implemented. The largest problem with the site was how to accommodate a major thematic architectural landmark in the triangular space where the Suzhou River and the Huangpu River merge. If we weren't able to cope with this, it would have to be merely a small-scale sculpture. When we were considering this, it occurred to us that the memorial should be a pearl of Shanghai's mother rivers, the Suzhou and the Huangpu. With this idea of a pearl, our thinking suddenly crystallized and we imagined a "round island". This idea of an island would not simply lie in idealized geography of the space and the contrast with the surroundings, but at the same time would place the commemorative tower in the prime tourist areas of Huangpu Park and the Bund. Taking a bird's-eye view from the air one can feel the overall beauty of the design concept!

As to the second project, it is a monument located in the Number 3 Plantation of the Yunnan Institute of Tropical Crops in Jinghong, the capital of the Xishuangbanna Dai Autonomous Prefecture. It consists of "Monument to the Inspection of Premier Zhou" and "Monument to the Meeting of the Premiers of China and Myanmar", which commemorates the friendship between the Chinese and Myanmar people. In bringing the work to completion Mr. Sima Quan and I were the designers. The framework of the "Friendship Monument" was inspired by the form of two hands held together. Four identical folding-ruler-shaped components were meshed together to form a "well" shape laying on the blue water of a pool, reflecting the willingness of the two countries to respect each other and live together peacefully, as well as the longstanding friendship between the Chinese and Burmese passed down from generation to generation. The "Friendship Monument" echoes the nearby "Inspection Monument", which stands vertically on a high platform. The monument is composed of a ring of four square-shaped components. The vertical layout uses a traditional style of Chinese stone monuments. The four components symbolize four rubber trees, expressing that the rubber industry is growing steadily like a broad-leaf tree. In addition, cornerstones were set up on both sides of the approach to the monument platform, and the story of the monuments made the overall atmosphere more harmonious.

Design starts from natural instinct, but is realized by skilled hands. Sometimes architects wait for inspiration to come, and when it comes they put it down on paper. In our time, architectural design drawings did not come from computer software, but from the pens of actual, vibrant people. This is a fundamentally human talent, and this talent extends each piece of paper so that what is on it can withstand the pressures of time.

江：我是学建筑出身的，从事编辑行业也快 20 年了，我对园林的理解是，中国园林一直作为中国传统文化的重要载体传达着一个民族对美、对自然、对生活的理解与追求。自然诗意的居住一直是中国人的居住理想。无论是老庄自然之道的产生，还是以表现自然美为主旨的山水诗、山水画和山水园林的出现、发展，都贯穿了人与自然和谐统一的哲学观念。这种观念在今天的风景园林景观设计中仍然是极其可贵。向中国传统园林学习，贵在学其神韵，而非简单的形式抄袭。您在前面也强调了这一点，那您对中国传统造园方法和原则有什么见解吗？

Jiang: My major was architecture, and I have done editorial work for almost 20 years. My understanding of gardens is that Chinese gardens have always been an important vehicle for transmitting traditional Chinese culture, conveying the country's understanding and pursuit of beauty, nature, and life itself. The poetry of nature has always been an ideal fit for the Chinese people. Whether it is the creation of the "natural way" of Lao Tzu and Zhuangzi, or the emergence and development of landscape poems, paintings, and gardens as ways of expressing natural beauty, they are all different locations on the spectrum of the philosophical concept of harmony between human and nature. In today's landscape-garden design this idea is still very precious. Learning from Chinese traditional gardens, we are learning their delicate beauty, instead of simply learning to copy. You also emphasized this point earlier. Do you have any opinions on traditional Chinese gardening methods and principles?

张： 过去人们造房多根据开间进深的格局，由营造匠人按照规定尺寸进行建造，然而建一个好的园子就不那么单一和固定了。能主事的人（即当时的园林设计师——文人或画家等）往往要依据地形、环境、想象、意境，因势利导进行现场构思，故很大程度上有着相对的灵活性、随意性。尤其是园林中的环境设计，仅用图纸很难表现得尽善尽美，要靠现场条件进行发挥和处理。就以园林中假山为例，在图纸中只能定个布局原则和大致的轮廓，堆砌时要根据所提供石料的形状妙思巧干、扬长避短，做假成真地创造出气韵生动的山势。这颇似一位雕刻高手，通过观察揣摩石料的色泽和纹理，以其高超的技艺雕琢出精妙绝伦的艺术精品。相比较一般的建筑设计，中国园林设计的文化属性更甚，其本质是园林设计者运用工程技术手段，艺术地演示一种文化现象。所谓的感觉，即园林设计者内在的艺术修养之外化。

Zhang: In the past, people usually designed houses assuming the standard width and depth of a room, and the builders built them according to the required dimensions. However, building a good garden is not so simple or formulaic. The person in charge (that is, the garden designer, a painter or some such) often had to design the scene according to the terrain or environment and his own imagination and artistic conception. Especially with respect to the garden's environmental design, it is difficult to put down a perfect plan on paper; one must take account of and play around based on the actual site conditions. For example, take the element of the rockeries in the garden. In the drawings, one can only set layout principles and draft approximate outlines. When actually stacking rocks one needs to, based on the shape of the stones provided, be smart and imaginative. This capacity is quite similar to that in the sculptor's hand. By observing the color and texture of the stone to be sculpted and using the superb skills, the sculpter crafts exquisite works of art. And compared to architectural design overall, Chinese garden design has more cultural attributes. In essence, garden designers use engineering techniques to illustrate a cultural phenomenon artistically. Any so-called "feeling" is the externalization of the garden artist's internal artistic cultivation.

江：我们知道中国传统园林具有极其鲜明的空间特征，综合运用借景、对景、框景等设计手法，创造出回环曲折、“山重水复疑无路，柳暗花明又一村”的空间序列。在重点部位则精心安排视点，组织视线，无论动观静赏都有别样风景。在这过程中有很多灵活性，要因势利导进行现场构思，您能给我们举例说明在潜园建造过程中出现的一些因环境变化而做的调整吗？

Jiang: We know that traditional Chinese gardens have very distinctive spatial characteristics, and comprehensively use design techniques such as borrowed and contrasting scenery, contrasting landscapes, and structural landscaping to create a spatial sequence of winding loops — when there seems to be no way out, one will suddenly find a way [here citing a Song Dynasty poem, later raised by the author in the text]. In the key parts, one should carefully arrange the lines of sight, and organize the scenery, so as to make it look different when viewing it from different vantage points. Doing things this way gives one a lot of flexibility. To take advantage of the ideas the designer has to be on-site. Can you give us examples of some adjustments made during the construction of the Chinese garden in Germany due to environmental changes?

张：因形就势，自然而然，“虽由人作，宛自天开”。换句话说，就是要与周边环境和地形和谐，并且不要“下手过重”。我举三个例子：

其一，外墙的处理。潜园建在鲁尔大学植物园中，可谓园中园。其格调源于江南园林，但不能移植，忌雷同，要“废前法”、立新意，外墙的处理也以此为准。高墙深院的苏州园林在这里就不那么相宜了。原设计是根据第一次选址的特点进行构思的，园墙建造在有一层高差的挡土墙上，为了构图的完美，在大台阶起步处的墙上设洞及垂花挑檐，此时进厅之后墙面为避重复，就不作重点处理。但当潜园移入植物园新址中，有一条园路正对园墙的另一端，此时将园墙重点处理的圆洞挑檐作相应的移动，使其成为一个对景，而远离此处的门厅后墙增设扇形景窗，以形成一个新的完美构图。

其二，圆形砖雕的重新安排。潜园大门之对面，水池中立一影壁，原设计中有一圆形砖雕起到一定的障景作用，而在新的园址周围树木郁郁葱葱，春绿秋黄，根据“俗则屏之，嘉则收之”的原则，若仍在此装置砖雕，反而掩盖了天成的美景，为此撤去砖雕，需另寻他用，苦思良法，望物尽其能。最后我们选择把它镶入折廊墙上的办法。廊中的墙面正需补白，将砖雕分成四块，并着意加工砍凿成残片，宛若出土文物镶嵌廊中，顿时使这一新建园林增添了难得的历史气氛和文化内涵。

为了使光线透入廊内，在墙的边缘又增开了缝隙，将墙外的色彩也引入园中，这种做法也立刻得到德方的认同和赞许。此手法符合中国传统亦适合欧洲习惯。在意大利、德国，有些建筑也往往于重点部位或入口处镶嵌残柱、拱券等，以达到点缀和装饰的效果。

其三，“野渡无人舟自横”是唐朝诗人韦应物的名句，潜园内有一景即取其意，名为“无人野渡”——待渡处。原来设计是想以草顶、木架、树桩、地坪来表达一种原始的自然情趣，但当假山逐渐堆置成形时，颇感这儿似乎是山的余脉、岩的根系。而以石垒矶砌岸，时聚时散，点缀水中构成一幅建筑与自然相互结合的山水画。此处虽没有船停泊，但通过自然布局，就会使人联想到待渡，联想到似乎是自古遗留下来的天然码头，也达到了“此时无声胜有声”的境界。

以上几例可见园林设计只在二维的图纸上难以表达充分，要尽善尽美也只有在三维的实际环境中，通过观察及再思考，精心斟酌才能创造出恰当的景观来。◎

Zhang: Impact follows form, naturally. “Though artificially made, it appears to be nature's work.” [Note: this remark is from *The Garden Treatise* (*Yuanye*), a Ming book on Chinese gardens.] In other words, the garden should be in harmony with the surrounding natural scenery, and not be overdone. I can give three examples:

First, how to build the outer walls. At Ruhr University, we were designing an inner garden built inside a larger garden, which can be thought of as a park in this larger garden. The Chinese garden's style comes from the Jiangnan Garden in Suzhou, but the original style could not be transplanted. Duplication had to be avoided; the “laws” of the old garden had to be repealed, and new ones established. The high walls and deep courtyard of Suzhou Garden were not so suitable here. The original design was conceived according to the characteristics of the original site. There, the garden wall was built on top of a retaining wall of non-uniform height. Accordingly, in order to avoid duplication, there was no attempt to emphasize particular features in the interior. However, at the new site, on the other side of the garden wall there would already be a path. So the spaces in the garden wall were moved, creating contrast, and fan-shaped windows were added to the back wall of the foyer to create a new perfect beauty for a new place.

Second, rearranging round brick carvings. Opposite the gate of the Chinese garden, there is a screen wall in the pool. In the original design, a circular brick sculpture played a certain role in breaking up the scene, while the trees around the new garden site were lush, spring green and autumn yellow. According to the [gardening] principle of “separating the vulgar and absorbing the elegant” , if the brick carving were installed in the same place, it would conceal the ideally beautiful scenery. To remove the brick carving for this reason, you need to think hard about finding another perfect place for it and about good techniques, and hope to do your best. In the end we chose to put it on the wall of the gallery. The wall along the gallery was in need of “filler” objects. The brick carving was divided into four pieces, and then each object was cut up further, in the manner of unearthed relics, which immediately added a distinctive historical atmosphere and cultural overtone to this new garden. In order to let the light penetrate into the corridor, a gap was opened at the edge of the wall, and the colors seen outside the wall were also introduced into the garden. This approach was immediately recognized and praised by our German partners. This method is in line with Chinese tradition and also suits European habits. In Italy and Germany, some buildings are inlaid with residual columns and arches at key points at entrances for purposes of embellishment and decoration.

Third, "I pull myself across on an unmanned country ferry" is a famous line by the Tang Dynasty poet Wei Yingwu [from a poem called "Chuzhou West Stream"]. There is a scene based on this line in this Chinese garden called "Yedu"(Unmanned Wild Crossing) , which is a place to be crossed. The original plan was to use the thatched roof, wooden frames, tree stumps, and the ground to convey a basic natural pleasure, but when the rock-garden portion was gradually stacked and formed, it felt like it was a remnant of a mountain with a root system of rocks sometimes clustered together and sometimes scattered, so that the water formed a landscape painting combining architecture and nature. Although there is no boat there, through the natural layout people will think of waiting to cross, and that it seems to be a natural wharf left over from ancient times.

The above examples show that garden design is difficult to express fully in two-dimensional drawings. To achieve the ideal in an actual three-dimensional environment, one needs to observe and rethink. This kind of careful consideration can create the appropriate landscape.

江：中国园林艺术和西方园林艺术是世界园林艺术的两大流派。风格迥异，表现形式也迥然不同。如西方人喜好雕塑，在园林中有着众多的雕塑。而中国人却喜欢在园内堆假山，中国人看树赏花看姿态，不讲求品种，赏花只赏一朵，不求数量，而西方人讲求的是品种、数量，以及各种花在植坛中编排组合的图案，他们欣赏的是图案美。是这样吗？

Jiang: Chinese and Western garden art are two substantial schools of thought. Their styles and expressive methods are of course different. For example, Westerners like sculpture, and so their gardens tend to have them. In contrast, Chinese people like to place rocks in their gardens, and to look at and admire trees, flowers and their arrangement. They don't stress variety; when they admire flowers they admire them one at a time. They don't emphasize quantity, while Westerners emphasize these things, so that the beauty is in all the flowers arranged together. Is that correct?

张：确实如你所讲的，在设计的时候，建筑是硬件，软件就是植物配置。植物配置我主要跟德国的植物园主任交流，在中国园林中看不到那么丰富的颜色。以苏州园林来说，基本以绿为主，没有那么多明显的颜色。而对德国人来讲，植物丰富的颜色是园林很耀眼的一点。园中种植过多、过杂，千姿百态，喧宾夺主。在和德方交流后，他们还是很尊重设计师的意见的。园林植物配置，并无固定模式，但总是取法自然，因地制宜地进行，做到"虽由人作，宛自天开"。潜园的植物配置还是比较独具特色的：春至草如碧丝，桑低绿枝；夏日莲叶无穷碧，映日荷花别样红；秋来红橙黄绿青蓝紫，最是一年斑斓时；冬临潜园，银装素裹，忽如一夜春风来，千树万树梨花开。一年四季，景色变换，期间流转，美轮美奂。

Zhang: Indeed, as you said, when designing, structures are the hardware and arrangement of plants is the software. With respect to plant arrangement I primarily engaged with the German director of the outer, larger botanical garden, and I never saw such rich colors in Chinese gardens. As far as the Suzhou Garden is concerned, it is mainly green; there are not so many bright colors. For the Germans, rich plant colors are a dazzling feature. But for me there were too many plants, and too many kinds of them. With this kind of over-satiation, a single element might distract from the whole arrangement. But after communicating with the German side, they still respected my opinions as the designer. There is no single, fixed pattern for the arrangement of garden plants, rather it is always natural and adapted to local conditions, so that, again, "though artificially made, it appears to be nature's work". The plant configuration in the inner garden is quite distinctive: In the spring, the branches are green and the grass sprouts [a Chinese saying from a poem by another Tang poet, in which the arrangement of the natural scenery a woman is looking at makes her think of her absent husband]. In the summer, lotus leaves to the horizon, boundless green, the sun glows on lotus buds, peerless red [from a poem of the Song Dynasty, "At Dawn, See off Lin Zifang at Pure Benevolence Temple"]. Autumn brings red, orange, yellow, green, blue and purple, making it the most colorful time of the year. When winter arrives the inner garden is covered in silver, as if suddenly the spring comes, and thousands of pear trees bloom. The scenery changes throughout the year, flowing beautifully.

江：潜园设计于 1987 年，1990 年 11 月落成。这是一个独特的中国园，远渡重洋，坐落于德国鲁尔大学校园内，设计精巧，深得中国园林精髓，在细节处又有现代的创意，是不可多得的近现代古典园林的佳作。在造园过程中一定也遇到过一些困难吧？

Jiang: This chinese garden was designed in 1987 and completed in November 1990. This is a unique Chinese garden across the ocean. It is located on the campus of Ruhr University, in Germany. Have you encountered some difficulties during the construction process?

张：在远隔千山万水的欧洲、与我们传统习惯各异的德国建造一座中国园，必然会碰到许多预料不到的情况。又因园址几度变动，根据新情况作相应的调整更显必要了。中国造园多取意于诗画，现场建造需多变化。而德国人又以严谨著称，仅举一例可见一斑。潜园主厅有临水墙基，我们习惯采用毛石砌筑，设计时在图纸上仅凭个人艺术感觉随意划分石块大小，施工时则视石料大小配合，达到一定要求即可。德方则不然，他们将毛石墙基的原设计图放大，并将划分的石块编号排队，送工厂逐块加工成型，运到现场再“对号入座”，予以固定，由此可见，不同的传统习惯对同一件事的处理方法差异如此之大。潜园中的建筑构件都是从中国海运而来。其实，设计和建造过程内外争论也不少，如建议摆上中国石狮子、金鱼缸或在 20 米白墙上做花窗等，但这些并不符合《桃花源记》中的立意，也就作罢了。

Zhang: In building a Chinese garden in Europe, partitioned as it is by many mountains and rivers, and in particular in Germany, whose traditional habits are different from our own, one will inevitably encounter many unexpected situations. Because the site changed several times, it was even more necessary to make the corresponding adjustments. Chinese gardeners are more inspired by poetry and painting, and the site construction thus needed even more changes. But the Germans are known for their strict precision, which one example will suffice to illustrate. The main hall of the Chinese garden has a waterfront. We are accustomed to using coarse rock. When designing, we can divide the size of the stones at will based on personal artistic preference. During construction, we can meet certain requirements based on the size of the stones. The Germans were different. They scaled the masonry wall foundation from that in the original design, and sent each rock sequentially to the workshop for processing and shaping. The rocks were then transported to the site and "sat down" to be finalized. This illustrates how customs can have such large effects on a particular task. The building components in the garden were all shipped from China. In fact, there were also many internal and external disputes during the design and construction process, such as over whether to add a Chinese stone lion, a goldfish tank, to put a flower window on a white wall of 20 meters, etc., but these did not conform to the original intention of the story "Peach Blossom Spring". In the end there ideas put into practice were not.

江：对于中国人来说，像潜园这样的中国之美展示在外国人眼前是一件让人觉得非常幸福的事。作为中德双方学术友好的象征，潜园的设计也倾注了您的智慧和心力。谈谈您在德国设计、建造潜园过程中的一些体会吧？

Jang: For you as a Chinese person, to be able to present some of the beauty of Chinese culture directly to foreigners through something like the Chinese garden must be very satisfying. As an emblem of German-Chinese academic friendship, this garden required you to pour in your wisdom and dedication. Could you talk a bit about your experiences while designing and building this garden?

张：自古艺术推陈出新，园林也是如此。在如今全球化的语境中，风物长宜放眼量。中国园林设计历史悠久，是世界艺术之苑中的一株奇葩，其本身蕴含着深厚的文化属性。在国外造园，更是一种中外文化间的交流与融合，不能一味模清仿宋、因循守旧，而要开拓思维，拓宽视野，因地制宜，根据实地条件再现一个理想的境界。

Zhang: Art has evolved since ancient times, and so too have gardens. In today's globalized context, it's wise to look at the situation. Chinese garden design has a long history and is a wonderful contribution to the broader art of gardening. It is part of culture in a profound way. Building gardens in foreign countries is a kind of exchange and integration between Chinese and foreign cultures. We cannot simply follow the old and imitate the practices during the Song or the Qing Dynasties, but we must open up our thinking, broaden our horizons, adapt to local conditions, be willing to let our tradition grow in new soil, and based on actual conditions create something perfect.

潜园的诞生
园林布局
粉墙疏影
入口——韵染阶前
进厅与楹联
少就是多
游廊漫话
屋舍俨然
园中一隅
主厅参差
砖雕似古
步入幽径
即兴小景
野渡
草棚风雨

画谈潜园
中国园林在德国

张振山 著
[美] 欧思博 译

A Chinese Garden in Germany

By Zhang Zhenshan
Evan Osborne Tr.

同济大学出版社
TONGJI UNIVERSITY PRESS

画谈潜园

中国园林
在德国

张振山 著
[美] 欧思博 译

By Zhang Zhenshan
Evan Osborne Tr.

A Chinese Garden
in Germany

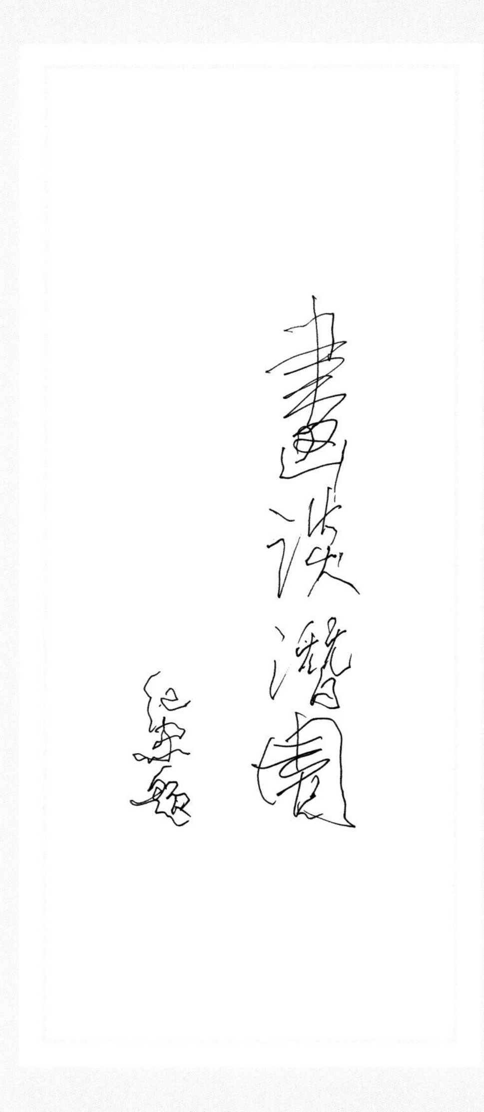

同济大学建筑与城市规划学院名誉院长
冯纪忠教授题字

目录

Contents

序言 1

埃尔马 · 魏勒教授
波鸿鲁尔大学前校长
2015 年春于波鸿

在潜园这座意为世外桃源的屋檐瓦当上，人们可以看到“五福万寿”的字样。在不乏魅力的波鸿鲁尔大学，潜园特具异彩。它是远在上海的同济大学作为合作院校于 1990 年赠送给鲁尔大学建校 25 周年的校庆礼物，也是两校多年持久友谊的象征。它是一所向公众开放的园林，是大学植物园内一处让人流连的所在。

潜园无与伦比的魅力源于它的原汁原味，因为它既非仿古，又非复制，而是中国原创。它是由张振山教授——一位中国的园林艺术学者设计并参与建造的。而中国的专业工匠用在中国预先制作好的正宗建筑部件， 建成了这座园林。它是一块独一无二的瑰宝！

潜园大概是波鸿最受欢迎的一处摄影景致，埃伯哈德 · 科赫先生还曾出版了一本精美的潜园影集。尽管如此，我们至今缺少一本介绍潜园的设计、建造过程以及综合描述其构造方案的第一手史料文献。

让我们欣喜的是，眼前这本为庆祝我校 50 年校庆而出版的双语画册，不仅可以填补上述空白，而且，基于该画册直接出自张教授之手，所以，喜爱这座中国园的朋友在阅读时还可以获得有关潜园的纪实描述。

感谢所有为本书的出版作出贡献的人士！

祝大家阅读愉快！让我们相会在潜园！

Prof. Elmar Weiler
Former President, Ruhr University Bochum
Bochum, Spring 2015

On top of the eaves on the roof of the Qianyuan Chinese Garden, whose name means "mythical paradise", people can see the words "Five Blessings and a Long Life" [a traditional Chinese blessing often found on buildings or smaller objects]. At the charming Ruhr University in Bochum, this inner garden located in the university's larger botanical garden is particularly colorful. It was a gift from Tongji University in faraway Shanghai given to commemorate the 25th anniversaryof the opening of Ruhr University in 1990, and it is also a symbol of long-term friendship between the two universities. The garden is open to the public, and it is a place that makes people touring in the larger botanical garden.

The unparalleled charm of the Chinese garden stems from its original flavor, because it is neither an antique nor a copy, but genuinely Chinese. Professor Zhang Zhenshan, a Chinese garden-art scholar, designed and helped build it. And authentic building components made in advance in China were used in constructing it. It is a unique gem!

The garden is probably one of the most popular places to take pictures in Bochum. Mr. Eberhard Koch has also published a beautiful photo album of it. Nonetheless, we have so far lacked a first-hand historical document to introduce the design and construction plan and process.

What makes me happiest is that this bilingual picture book published to celebrate the 50th anniversary of our school can not only fill the above gaps but is also written by Professor Zhang himself. And friends who love this Chinese garden can also get a documentary description of it while reading.

Thank you to all who contributed to the publication of this book!

I wish you all pleasant reading! Let's meet in the garden!

序言 2

克劳斯 · 科施通教授
德国国家级建筑大师、教授
德国欧博迈亚设计咨询有限公司原总建筑师
2014 年秋于慕尼黑

是荣幸，是信任，是感激，同时也是一种享受，让我为我尊敬的老朋友，张振山教授的新书作序。

张振山先生是上海著名的同济大学的教授。他在同济大学教授建筑学、城市规划以及风景园林等专业的课程。张教授桃李满天下，学生遍及海内外。

张教授与我相识于 15 年前的一次建筑设计竞赛——一个为柏林的中国驻德大使官邸举行的设计竞赛。他当时是该项目的中国顾问，而我则是慕尼黑的欧博迈亚公司的中国区首席建筑师。通过这次合作，我们迅速地建立起一种真诚的、建设性的友谊关系，或许是因为我们志趣相投吧。

由此开始，我们之间架起了一座中德文化交流的桥梁。我们在中国的项目中相互支持，共同探讨哲学的、社会政治的以及当代的诸多问题。我们相互交换自己在规划、哲学、未来城市发展的基本问题等领域的基本思想。

1990 年的天时、地利、人和，让张教授在鲁尔大学的校园内实现了他的中国园林。这对德国来说是一件喜事：德国自此拥有了一个由中国著名专家设计的中式园林。它位于大学校园内，各个专业的师生以及普通的市民，都可以学习、欣赏到这样的一种文化、这样的一种美。这个园林，其实也可以看作是一种欧洲传统的延续：自 16 世纪始，欧洲人开始在艺术、建筑、园林文化、日常用具、饮食文化等方面，受到中国文化的影响。

鲁尔大学的这座中国园林，自它开园之日起，便获得社会广泛关注，好评如潮，引起公众极大的兴趣。现在，张教授用他当年在设计时体现出来的热情，以及他对中国神话与哲学的孜孜追求，完成了本书的著写。我们无需解释，在潜园就如在园林城市苏州，身临其境地参观了一座中国园林。因为园林本身就叙说着一切，它们遵循着几百年来的建筑哲学、设计、格局和经验，一切都深深的根植于“人天合一”的建筑理念中。园中的流水、假山石、树木、灌木、花草以及建筑和回廊等建筑元素，相辅相成，呈现出一种和谐动人的节奏变化。人们似乎也可以把这些园林艺术称之为“园林交响曲”。

本书独特的魅力还在于图片顺序与内容的相互交叉。张教授让我们“参与”了他当初设计园林时的历程——图纸、手绘草图、评论、详情说明等原始资料，按顺序排列，与精美的园林实景照片配合在一起，相得益彰，以致读者在看完这本不同寻常的书之后，留下一个强烈的印象：不仅畅游了一次中国园林，同时也在作者专业角度的引领下，以迷人的方式，被带入一个中国园林艺术和中国哲学的秘境。

FOREWORD 2

Prof. Klaus Kohlstrung
German master architect and professor
Former Chief Architect, Obermeyer Design Consulting
Munich, Autumn 2014

It is an honor, a trust, something I am grateful for, and also a particular pleasure to write this foreword to the new book by my old respected friend, Professor Zhang Zhenshan.

Mr. Zhang Zhenshan is a professor at Shanghai's famous Tongji University, where he teaches courses in architecture, landscape architecture and urban planning. Professor Zhang's former students can be found worldwide.

Professor Zhang and I met at an architectural design competition 15 years ago — to design the official residence of the Chinese ambassador in Berlin. He was a Chinese consultant for the project at the time, and I was the Chief Architect for China Office of Obermeyer in Munich. Through this cooperation, we quickly established a sincere and productive friendship, which may be due to the spiritual kinship between us.

Since then, we have built bridges for Sino-German cultural exchanges. We support each other on projects in China and discuss philosophical, sociopolitical, and contemporary questions. We exchange the basic ideas we have in the fields of planning and philosophy, and basic issues of future urban development.

In 1990, a propitious time, the right location, and harmony among the right people allowed Professor Zhang to realize his Chinese garden on the campus of Ruhr University. This was a happy event for Germany; the country now owns a Chinese garden designed by a famous Chinese expert. It is located on the university campus. Teachers and students of all majors, as well as ordinary citizens, can learn and appreciate such culture and beauty. This garden can also be regarded as a continuation of a European tradition: since the 16th century, Europeans have been influenced by Chinese culture with respect to art, architecture, goods used in daily life, and food and garden culture.

Ruhr University's Chinese garden from the moment it opened has received a lot of attention, been admired by many, and generated much interest among the public. Now, Prof. Zhang has been able to use the enthusiasm he felt when designing the garden, as well as his longstanding pursuit of Chinese legends and philosophy, to finish the writing and editing of this book. Just as with the "garden city" of Suzhou, we need not explain a Chinese garden we are visiting. The gardens themselves tell everything — the architectural philosophy, design, layout, and experience of hundreds of years are all deeply rooted in the architectural concept of "the union of human and heaven". The flowing water, rockery, trees, shrubs, flowers, and architectural elements such as buildings and gallery complement each other and present a harmonious and moving rhythm. People can call such garden art a "garden symphony".

The unique charm of this book lies in the intersection of the graphics and the verbal content. Professor Zhang lets us "participate" in the process of his original garden design: original materials such as drawings, hand-drawn sketches, comments, and detailed descriptions, arranged in order, are complemented with exquisite photos of the garden. So after finishing this unusual book, the reader is left with a very strong impression: not only getting a tour of Chinese gardens, but also led in a fascinating way by the author's professional perspective into the secret realm of Chinese garden art, and Chinese philosophy.

前言

“潜园”设计于1987年，建造于1990年，至今已有二十多年了。起初，我未曾有写书的想法，因为写书对我来说非属易事，所以在施工过程中拍的许多照片，随着时光的流逝也就慢慢地失散了，手中只留了一些成果照作为纪念，仅此而已！多年前有些朋友看了照片，都鼓励我将其整理成册，后来说的人多了，才渐渐地萌生写书的念头；但又甚感资料太少，潜园又远在德国，相隔万里，鞭长莫及。有幸的是2007年得到德国朋友拜尔曼建筑师寄来数十张非常精彩的潜园照片，因此，对写书之事才增加了信心和兴趣。借此我把设计和建造过程中的体会、心得整理出来与友人交流，供大家参考，也希望能再为学生讲一堂园林设计课，这将是让我非常愉快的一件事。

万分荣幸地请到了冯纪忠先生为本书题写书名。冯先生是我最敬慕的前辈，他不仅是同济大学建筑学科的领衔人物，而且被公认为是中国现代建筑设计思想的倡导者和奠基人。他学识渊博、纵横中西，是我国建筑界的一代宗师。早在20世纪60年代初，他倡导的“空间原理”教学，是对建筑教育的一大贡献。上海松江的方塔园是冯先生的一大杰作，该园设计在世界建筑师大会优秀设计展上，荣获50个优秀设计作品中唯一的园林设计奖。2008年举办的首届中国建筑传媒奖评选，全国建筑界经过公开、公平的投票评选，“杰出成就奖”就颁发给了冯纪忠先生。这些都应验了中国的一句古话：“桃李不言，下自成蹊。”

谢谢冯纪忠先生，在他94岁高龄时为本书题字！

为本书写序言的是一位德国朋友科施通教授，一位德国建筑大师。我极为高兴的是，他在百忙之中能抽出时间写序，而且几经修订，甚是认真。科施通先生是德国欧博迈亚设计咨询有限公司原总建筑师，我们不但是相识多年的老朋友，而且我的学生和校友也曾是他的助手和伙伴，所以我们之间有着多层面的立体式的相识。他不单对中国建设有着杰出的贡献，而且对中国的文化也有着浓厚的兴趣。他对中国太熟悉了，用他的话说：“每天早晨醒来，先要想一下：是在北京？上海？还是在慕尼黑？”中国已成为他的第二故乡了。

潜园建成二十多年后，由他为本书写序，是又一次中德两国同行的合作，这应该是快乐的延续，是一次很有意义的合作，衷心感谢科施通教授为《画谈潜园》撰写序言。

正当该书出版之际，波鸿鲁尔大学前校长埃尔马·魏勒教授接受了我们的请求，为《画谈潜园》作序。这使我们感到既亲切又荣幸，千言万语难以尽表，还是借用魏勒校长的一句话，作为前言的结束语：“让我们相会在潜园吧！”

张振山
2015年春于上海

PREFACE

Zhang Zhenshan
Shanghai, Spring 2015

The Qianyuan Garden was designed in 1987 and built in 1990, and it has been more than two decades since then. Initially, I did not think of writing a book, because for me writing a book was not easy. As a result, many photos were taken during the construction process gradually disappeared with time, leaving only a few in my hands. What remains should only be taken as a memorial, nothing more! But some friends saw the photos many years later, and they encouraged me to organize them into a book. Later, more people said the same, and I gradually began to think of writing it. But I felt that there was too little information, and the garden was in Germany, thousands of miles away, and I had no way to return. Fortunately, in 2007, I received dozens of wonderful photos of the garden from a German friend Beilmann (an architect). This gave me sufficient confidence and motivation to write this book. In writing it I have arranged my own experiences during the design and construction process and been able to communicate them to friends for their reference. I also hope that this book will be a garden-design lesson for the students, which would give me great pleasure.

Thank you Mr. Feng Jizhong for inscribing this book at the age of 94!

The foreword for this book was written by a German friend, Professor Kohlstrung, a master German architect. I am extremely pleased that he could take the time to write this given his busy schedule, even through several revisions; he was truly dedicated. Mr. Kohlstrung is the former chief architect of Obermeyer Design Consulting Co.,Ltd in Germany. Not only have we been friends for many years, but my students and alumni have gone on to be his assistants and partners, so we have a multi-dimensional relationship. He has not only made outstanding contributions to construction in China but also has a strong interest in Chinese culture. Indeed, he is perhaps too familiar with China; in his words: he "wakes up every morning, and first wonders, is this Beijing? Shanghai? Or Munich?" China has become his second home.

Twenty years after the completion of the Qianyuan Garden, his writing the foreword here is another example of cooperation between China and Germany. This should lead to continued good relations and is a truly significant cooperative endeavor. My heartfelt thanks to Prof. Kohlstrung for writing the foreword to *A Chinese Garden in Germany*.

Just as the book was getting ready to be published, former President Elmar Weiler of Ruhr University Bochum accepted our request and wrote another foreword for *A Chinese Garden in Germany*. No words would be sufficient to say how delighted and honored we are, so I will borrow the words of President Weiler at the end of his foreword: "Let's meet in the garden!"

潜园的诞生

Birth of the Garden

01

林木苍劲、落叶纷陈，烘托于粉墙黛瓦的建筑群前，在遥远的异国他乡，更会让人感到新奇、神秘，在那幽深的高墙大院内，一定还隐匿着某个古老的故事！

故事还得从德国鲁尔大学说起。鲁尔大学位于波鸿市郊，是德国著名的高等学府，与中国的同济大学是最早的兄弟友好院校。作为鲁尔大学 25 周年华诞献礼，并得益于波鸿市政府和当地储蓄银行文化基金会的大力支持和资助，他们共同商议拟定在校园内建造一处中国园林，让中国园林艺术永远留存在那儿！

1987 年我在德国慕尼黑学习期间，接到同济大学高廷耀校长的电报（当时还没有电子邮箱），要我去鲁尔大学商议中国园林的设计事宜。三天后，我到了当时联邦德国的首都波恩，中国大使馆教育处的赵其昌教授带我去了鲁尔大学，并见到了当时的校长沃尔夫冈 · 马斯贝格教授。最初他们把园址确定在教学楼旁边一块不大的空地上。我在那儿停留了两天，根据他们的想法，试探性地构思了草图。当我回到慕尼黑一周后，就接到鲁尔大学校办的邀请，得知校长办公会议已同意了方案草图的初步构想。这时，我才对这个项目有了落实的感觉，随后，我约请了一起在德学习的龙永龄和郑孝正两位老师，北上波鸿鲁尔大学，具体设计这处即将建在德国的中国园林。

因园址最初确定在高层林立的教学楼旁，本着《园冶》中“俗则屏之，嘉则收之”的精神，设计时就要充分考虑到将游人的视线避开外界高楼的视觉干扰。这很自然使我们联想到《桃花源记》的“避秦时乱”，联想到作者陶潜（陶渊明），“潜园”因此而得名。“桃花源”的基调，也顺理成章地成为该园的立意及内容取舍的依据，建筑造型简朴素雅，民居格调，总体布局遵循自然，环境宜恬淡清幽。可谓集古朴之风韵，寓淡泊之情怀。

1990 年施工时，园址改至绿树成荫的植物园中。但此前因建筑布局及构架均大局已定，我们只能根据新址的现场情况做些调整，使潜园得到一些借景机会。

潜园的主体建筑由无锡园林古建公司承建，并赴德国现场施工，先后由当地有关专业公司和德国工程师克吕格尔配合，共同施工。我和司马铨老师负责现场指导。施工时间甚短，1990 年 5 月正式开工建造，同年 11 月 29 日建成并举行了揭幕式。

居士高踪何处寻，
居然城市有山林。

《园冶》

《园冶》，计成所撰，体系齐全、内容丰富，系统阐述了如何造园的技法与理论，是中国第一部全面总结造园原则与方法的著作，涉及园林创作的各个方面。如此重要的园林建筑著作，自明末刊行后的三百年间，一直寂寂无闻，除李渔《闲情偶寄》中有一语道及，此外未见著录。究其原因，是因为甲戌本乃阮大铖资助刊行，出版时又有阮氏序言。阮大铖为人反复，品格低下，留下百世骂名，《园冶》也受池鱼之殃，清代甚至一度列为禁书。大约在清中期，《园冶》流传至日本，日本人极为尊崇，奉为「世界最古之造园学专著」（日本造园名家本多静六博士语），屡加翻刻。在日本国发现的《夺天工》，一直以来被认为是日本人对《园冶》的翻刻本，近来有学者考证此本实则清中期已有，系自中国传入。《园冶》在中国大放光芒已经到了二十世纪三十年代，中国营造学社创办人朱启钤，将北平图书馆发现的《园冶》残本，补成三卷，收入「喜咏轩丛书」，后又将其与东京内阁文库的明安庆阮氏刻本《园冶》对照整理，于一九三二年由营造学社出版《园冶》三卷本。此后又历经陈植、陈从周等人诠释和宣传，《园冶》的重要价值终于为人所知。

出自姜龙，董玉海主编，《扬州历代名著》，江苏：广陵书社，2017.02: 44-48.

The Garden Treatise

Written by Ji Cheng, the book has a complete system and is rich in content. It systematically explains the techniques and theories of how to build gardens. It is the first book in China to comprehensively summarize the principles and methods of gardening and covers all aspects of garden creation. Except for a reference in Li Yu's Xianqing OuJi [a book also written in the 17th century, but under the Qing Dynasty], this important work on garden architecture has been neglected for three hundred years since it was published at the end of the Ming Dynasty, and previously no records had been found. The reason for this is that the original work was sponsored by Ruan Dacheng [a Ming poet], and included a preface by him when it was published. Being politically unreliable, Ruan was seen as someone of low character, and left behind a century of infamy. *The Garden Treatise* was thus unjustifiably condemned, once even being listed as a banned book during the Qing Dynasty. Around the middle of the Qing Dynasty it spread to Japan, where it is revered and regarded as "the world's oldest monograph on gardening" (according to gardening master Dr. Seiroku Honda), and there have been numerous translations. A book called *Perfecting Nature* found in Japan has long been considered to be a Japanese translation of *The Garden Treatise*. Recently, some scholars have verified that this book was already available in the middle of the Qing Dynasty, and it was inherited from China. And by the 1930s *The Garden Treatise* was seen as a masterwork in China. Zhu Qiqian, the founder of the Society for the Study of Chinese Architecture, supplemented the incomplete book found in the Peking library into three rolumes, published it in the Xiyongxuan series, and then compared it with the Anqing Ruan's block-printed verison of it in Tokyo's "Cabinet Library" (the Naikaku Bunko). The Society for the Study of Chinese Architecture published the treatise in 1932 in three volumes. Since then, Chen Zhi, Chen Congzhou, and others have interpreted and publicized its important value.

From Jiang Long, Dong Yuhai, eds, *Yangzhou Famous Works*, Jiangsu Guangling Press, February 2017: 44-48.

水阁临风迥，莲舟隔浦寻。
红妆风四面，翠倚月中心。

如隐世之秘境，浮华室外，恰宁静致远，独享方圆。

Venerable trees with their many fallen leaves, in front of a building complex with a running white wall and black tiles. In a distant exotic country, people will regard such a scene as new, mysterious. Inside a deep high-walled courtyard, there certainly must be an ancient story hidden!

This story must start from Germany's Ruhr University. The university is in the suburbs of Bochum. It is a famous elite German university, and one of Tongji University's earliest sister schools. To serve as a 25th birthday gift for Ruhr University, and with strong support and funding from the Bochum city government and the cultural foundation of the local bank, it was jointly agreed to plan to build a Chinese garden on campus. And that Chinese garden art will always be there!

When I was studying in Munich, Germany in 1987, I received a telegram from President Gao Tingyao of Tongji University (there was no e-mail at the time), asking me to go to Ruhr University to discuss the design of a Chinese garden there. Three days later, I arrived in Bonn, then the capital of the Federal Republic of Germany. Professor Zhao Qichang of the Education Department of the Chinese Embassy took me to Ruhr University and I met the then-president, Professor Wolfgang Maßberg. Initially, they identified the site as a small, open space next to the school building. I stayed there for two days, and based on their ideas, tentatively drafted the sketch. When I returned to Munich a week later, I received an invitation from the university, and learned that the president had been in a meeting that had agreed on the preliminary idea of the draft plan. At this time, I had a sense that the project would be implemented. Later, I invited two teachers, Long Yongling and Zheng Xiaozheng, who were studying in Germany, and went north to Bochum Ruhr University. A specifically designed Chinese garden was going to be placed in Germany.

In line with the spirit of *The Garden Treatise*, because the park was originally set up next to high-rise teaching buildings, during design I had to take full account of avoiding these buildings disturbing visitors' line of sight. This naturally reminds the story "Peach Blossom Spring" and the author Tao Qian (also known as Tao Yuanming), through which the idea of a "hidden garden" became famous. The essence of this "Peach Blossom Spring" also logically became the basis of the park's concept, and its content selection. The architectural style is simple and elegant, in the style of residential buildings. The overall layout follows nature, and the environment is quiet. It can be described as having a simple kind of charm, unemotional.

During construction in 1990, the site was changed to a tree-lined garden. However, because the overall layout and structure of the building had been determined before, we could only make some adjustments based on conditions at the new site, but this meant that the Chinese garden presented some opportunities to borrow.

The main building in the hidden garden was built by the Wuxi Garden & Ancient Architecture Company, who went to Germany for on-site construction. It was built with the cooperation of local specialized companies and the German engineer Krüger. Mr. Sima Quan and I provided on-site guidance. Construction did not take very long. It officially started in May 1990, and it was completed and the garden unveiled on November 29 of the same year.

手绘轴测全景图

最初的园址

建筑木构架在国内加工试装

德中两国工人现场施工

园林布局

The Garden Layout

波鸿市位于德国西部，地处非常密集的城市群中。鲁尔大学就坐落在波鸿市南郊的一处坡地上，层层叠叠的大平台是步行人流的中心枢纽，它联系着各个教学大楼及生活服务设施（图书馆、大礼堂及餐厅等主要建筑物）。大平台下是随地形坡度变化的多层地下车库，而其林木葱郁的正前方是当地著名的植物园，潜园就建在植物园内（图中绿圈处）。

潜园占地很小，约 37 米 ×25 米，也就是俗话说的一亩三分地。全部建成后共分三部分：

（一）中间部分是园林的主体，小院之内占地一亩，园内以水面居中，游廊、厅、舍与自然山岩参差围合，环水而建。清澈的水面宛如明镜，四周景物均倒映水中，借以扩增小园的视觉空间。中国园林“贵在因借”，院内均可借到墙外树冠景色，园内园外相互渗透，春绿秋黄格外宜人，水上水下交相辉映，以期达到小中见大的艺术效果。

（二）主院以南是入口的前奏，它由进厅、大门、长方形的水池及巨大的粉墙组合而成。这是进入潜园的序幕，是园林内外的过渡景观。

（三）主院的北侧属二期工程，原为学术活动场所，后来德方拟辟为中国茶室。

Bochum is located in western Germany, in a very dense urban area. Ruhr University is located on a sloping field on its southern outskirts. The large, layered platform is the central hub for pedestrian flow. It connects various teaching and service buildings (libraries, auditoriums, restaurants and other major facilities). Beneath the big platform is a multi-level underground garage that changes with the terrain slope, and in front of its lush forest is the famous local botanical garden. The Chinese garden is built within this botanical garden (the green circle in the picture).

The Chinese garden occupies a small area, about 37 × 25 meters, a tiny piece of land. After it was completed, it was divided into three parts:

(1) The portion in the center is the primary part. The small courtyard occupies an area of one [Chinese] acre. There is a pool of water in the middle of this part. The gallery, halls, and buildings are surrounded by natural mountains and rocks. The clear water surface is like a mirror, and the surrounding scenery is reflected in the water, thereby increasing the visual range of the small garden. It is said of Chinese gardens that their magnificence lies in their borrowed space. The canopy scenery outside the walls does double duty in the courtyard; in spring green and in autumn yellow are particularly pleasant here. Water above and below achieves the artistic effect of seeing something bigger in something smaller.

(2) To the south of the main courtyard is the “prelude” to the entrance. It is a combination of an entrance hall, gate, rectangular pool and huge white wall. This prelude is a transitional space to the Chinese garden, located both inside and outside it.

(3) The north side of the main courtyard belongs to the second-phase of the project, which had originally been a place for academic activities. Later, the German side decided to set up a Chinese tea house.

潜园坐落在郁郁葱葱的植物园内

平面图

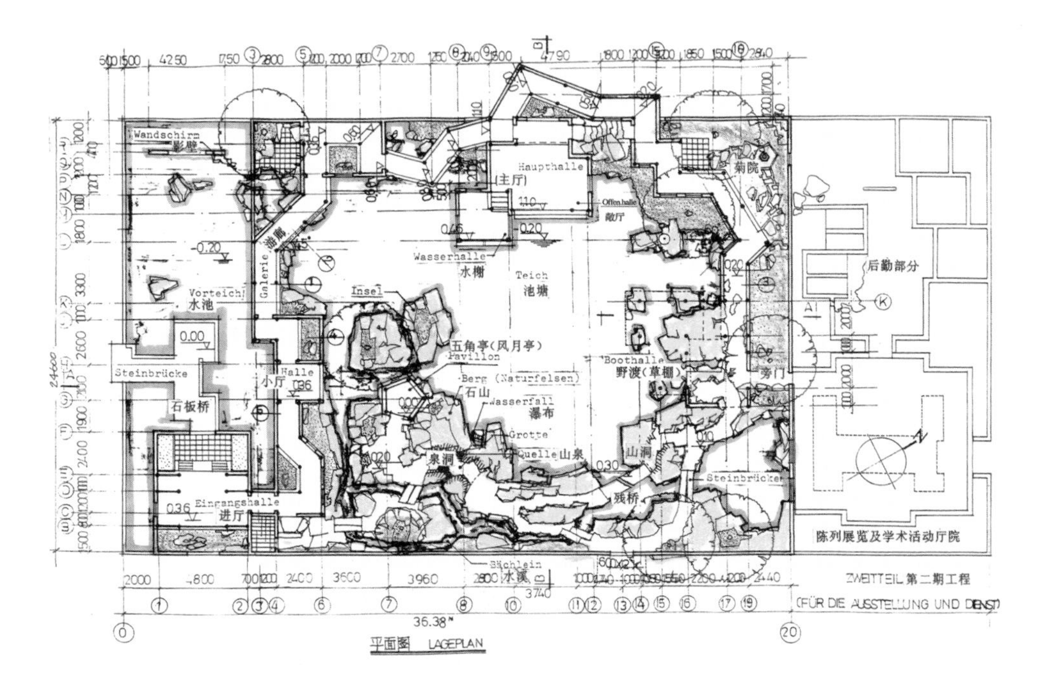

模型俯视

模型鸟瞰

粉墙疏影

White Wall, Scattered Shadows

028 – 029

“山重水复疑无路，柳暗花明又一村。”

进入植物园后，要穿过层层林地、草丛，方可到达潜园。有了前奏，才能渐入佳境，先抑后扬更能彰显园内景色的美。中国人的审美崇尚含蓄，就像吃饺子，在咬开之前不知道里面是荤抑或素？它不同于外国的匹萨饼，红红绿绿的，都摆在外面。

当游人驻足潜园门前，映入眼帘的是一处中国民居风格的深宅大院。若放眼四顾，则是秋风吹落叶，古树伴高墙，桥凌池上，水柔石刚，组成一幅既艳丽又质朴的空间画面。此是进入潜园之“序曲”。

既然是“世外桃源”，20米长的大墙宜封不宜透，宜屏也不宜堵。为此，在墙的左下角辟一水洞，园内园外水系相连，游鱼嬉戏穿梭其间。大墙内外似隔非隔、似通非通的处理手法，也体现着一定的含蓄之美。

白色的墙面酷似一处巨幅的“水墨丹青”。千姿百态的树影，展示给我们的是应时变化、如梦如幻的天地之画。偶逢微风徐徐阳光灿烂的日子，池面水波潋滟，墙面粼光闪耀，与树影交织出气势愕然、似幻似奇的光影画卷。

有人说，这个墙面称得上是“不着一字，尽得风流”，我心里说：“过奖了！”

“When the water and mountains seem to provide no way out, the trees and flowers can lead the way.” [Chinese saying about there always being a way out of any difficulty]

After entering the botanical garden, one must pass through layers of woodland and grass to reach the Chinese garden. The “prelude” space is needed to gradually enter the most beautiful part. The Chinese people believe in an implicit idea of aesthetics, in a sense like eating dumplings. Before biting into it, I don’t know whether inside there is meat or vegetables. It is different from pizza, where everything is placed on top.

When visitors stop in front of the garden, a deep courtyard in the style of Chinese folk houses comes into view. If you look around, you see the autumn wind blowing leaves, ancient trees with high walls, the bridge atop the pond, the water, and some soft stones. Together they make a beautiful and yet simple space. This is the “prelude” to the garden.

It is meant to be evocative of Peach Blossom Paradise, and so the 20-meter-long wall should be partially and not completely closed. To this end, a water gap was opened in the lower-left corner of the wall. The water flows inside and outside the park are connected, and fish swim through them. The inside and outside of the wall seem separate yet not separate, connected yet not connected, and so embody a certain subtle beauty.

The white wall seems like a huge “Chinese painting”. The many different tree shadows give us the impression of seasonal change, a dreamlike evocation of both heaven and earth. Occasionally, there are days with a gentle breeze, ripples in the pool and the wall gleams with light, interweaving with tree shadows. Then, it seems like a fantasy of a Chinese scroll painting filled with light and shadow.

There are people who say that for this wall “there are no words for this scene, only the romance”. And my heart tells me, “Too wonderful!”

红粉墙头花几树。落花片片和惊絮。

倚南墙，几回凝伫。
绿筠冉冉如故。

拂墙花影飘红。
微月辨帘栊。

入口——韵染阶前

Charm before the Entrance

鲁尔大学与同济大学的友谊标识

如果说，风花雪月阴柔之美，渗透着传统园林的主调；那么我们是否可以说，生动的光影变幻，给中国园林的设计增添了新意？

潜园的入口质朴典雅，古朴大方。从瑞士运购的黑色石板浮架水上，质地粗犷显得苍劲。石桥座栏围而不封，循势而立，以导人入园。大门、水池、石板相互衬托，浑然一体，景色宜人。

门前石岸，似曾系舟？院中虽与世人间隔，但也怡然自乐。在这儿四时更替景色迥然：有时是春光明媚疏影横斜；秋风落叶，门下阶前则色彩斑斓；冬雪之后，寂静素雅洁白如梦。遥忆当年，依稀故道尚留渔人足迹？问君：“此乃何方‘净土’？”答：“桃源仙境。”

If we speak of the pervasive yet gentle beauty of the wind, flowers, clouds, moon and sun as the main theme of traditional gardens, then can we say that changing the light and shadow will add new meaning to the design of Chinese gardens?

The entrance of the garden is simple and elegant, basic yet stylish. The black slate slabs brought from Switzerland float on the water, with their bold yet rough texture. The surrounding stone bridge is not closed, and is upright to guide people into the park. The gates, pools, and slate slabs are set off against each other, yet are a single entity, making for a very pleasant scene.

Does the stone bank in front of the door seem to be a boat? The courtyard is separated from the world, but also has a happy air. Here one can completely see the seasonal changes in scenery. The area is very bright in spring, and the leaves fall in autumn. In winter after it snows, it is elegant, silent, dreamlike. Looking back through time, would any ancient footprints of the fisherman still be there? The answer is “Yes, in the Peach Blossom Paradise.”

潜园入口

暮春和气应，白日照园林。

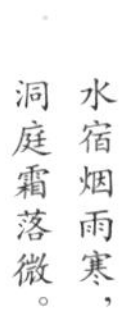

水宿烟雨寒，洞庭霜落微。

雪消门外千山绿，
花发江边二月晴。

隔牖风惊竹，
开门雪满山。

进厅与楹联

Entrance Hall, a Pair of Couplets

05

厅是入园后的第一景观，迎面的白色匾额“世外桃源”四个大字，点明了潜园的主题。两旁的楹联：上联是“为闻庐岳多真隐”，下联是“别有天地非人间”。在我的记忆中，这好像是从湖南桃花源牌坊上抄来的，字是请著名书画家郑孝同先生写的。白匾黑字素雅醒目，很有装饰性。前些年偶遇郑孝同先生谈及此事，他总觉得字写得不够好，甚为自谦。但从我们的眼光看，这已经是很棒了。作为一位书画家，相隔二十年的修养和眼力当然不可“同日而语”了。

题字和楹联，在中国园林中不仅是点缀，更有烘托和点景作用。潜园内各个建筑本应请诗人和文学家为其点题、缀文，但当时我们势单力薄，无力搬请“他山之石”，只能抽时间去沧浪亭选抄一二，算是沾了点文墨吧！没想到这一“抄”，竟抄出来一个很有意思的故事。

潜园内的主厅里，还有一幅对联是“清风明月本无价，远山近水皆有情”。郑孝正老师饶有兴致地对其查证，特摘录如下：“沧浪亭”始建于北宋，此前原是五代十国时期，吴越王贵戚的废园。北宋进士、著名文人苏舜钦以四万钱购得此处旧园，改建为“沧浪亭”，好友欧阳修题词中有“清风明月本无价，可惜只卖四万钱”的句子。斗转星移，一晃就是几百年，到了清道光七年（1827），苏州巡抚梁章钜在重修沧浪亭时，取欧阳修《沧浪亭》中的前半句和苏舜钦《过苏州》里“绿杨白露俱自得，近水远山皆有情”的后半句，组合为“清风明月本无价，近水远山皆有情”。因我当时并不知此是原句，为了念起来顺口我就改成，“远山近水皆有情”了。此名联流传各地，在我国许多风景名胜地都可看到它的传抄版，但它们的老祖宗就出自“沧浪亭”。而潜园的这幅对联应属得上是沧浪亭的直系第二代吧！德国的潜园与苏州的沧浪亭就有了某种直系“血缘”关系。

这是一个美好而真实的故事，巧亦巧矣，妙亦妙矣！真的要感谢郑孝正老师为潜园认证了这门“千古名亲”。

进厅后看到的匾额「世外桃源」，点明了潜园的主题。

白匾黑字、素雅醒目。

The hall is the first landscape one sees after entering the garden. A white plaque with four large black characters announces the Shiwai Taoyuan [Garden of the Eternal Peaches] and clearly points out the garden's theme. There are also poetic couplets on both sides: the right one [from a Song Dynasty poem by Su Shi] reads "I have heard Lushan Mountain is remote, and suitable for contemplation", and the left one [from the Tang Dynasty poem by Li Bai] reads "This is a world of its own, unlike the human world". I recall that this was copied from the Taohuayuan Archway in Hunan. The inscription was written by Mr. Zheng Xiaotong, famous calligrapher and painter. As a decoration, the black characters on white plaques are elegant and eye-catching. I met Mr. Zheng Xiaotong a few years ago and we talked about it. He was very modest, and always felt that the words were not painted properly. But as we look at it here, it is wonderful. Of course as a calligrapher and painter, two experiences looking at the same work after 20 years of self-cultivation cannot be compared with each other.

In Chinese gardens, all these inscriptions are not merely embellishments, they also serve as offsetting backdrop and focal points. Every component in the garden should use the works of poets and writers for its inspiration, but during construction, we didn't have much capacity to move such "stones from other mountains". We only had the time to choose one or two from the Canglang Pavilion. It was like having very little ink with which to write! I didn't expect this "copy" itself to tell a very interesting story.

In the main hall of the Chinese garden, there is another couplet, "The gentle breeze and the bright moon are priceless in their own right, and the faraway mountain and the nearby water both move us." Mr. Zheng Xiaozheng investigated these with interest. He has specifically noted that Canglang Pavilion was built during the Northern Song Dynasty. It was originally an abandoned garden of the King's relative Guiqi of Wuyue during the period of the Five Dynasties and Ten Kingdoms. The Northern Song Dynasty scholar and famous literati Su Shunqin bought the old garden there for 40,000 yuan, and rebuilt it as Canglang Pavilion. His friend Ouyang Xiu wrote, "The gentle breeze and the bright moon are priceless in their own right, and this spot is only worth 40,000 yuan". But after a few centuries of change, during the seventh year of the reign of Daoguang during the Qing Dynasty (1827), when the governor of Suzhou, Liang Zhangju, rebuilt the Canglang Pavilion, he took the first half-sentence of Ouyang Xiu's "Canglang Pavilion" and the second part of Su Shunqin's "Crossing Suzhou" — "The green poplar and the white dew are both satisfying, and the nearby water and the faraway mountain both move us", and created "The gentle breeze and the bright moon are priceless in their own right, and the nearby water and the faraway mountain both move us". Because at the time I didn't know this original sentence order, in order to render it easier to read I changed the order to "The faraway mountain and the nearby water both move us". This couplet name has spread everywhere, and it can be seen in many scenic spots in China, but it came from the Canglang Pavilion. And so this antithetical couplet [a more specific term for this type of Chinese poetry] in this garden should be thought of as Canglang Pavilion's second generation! The German Chinese garden has a direct "family relationship" with Suzhou's Canglang Pavilion.

This story is beautiful, and true. I really want to thank Mr. Zheng Xiaozheng for verifying this ancient, dear story for our garden here.

为闻庐岳多真隐，
别有天地非人间。

清风明月本无价，
远山近水皆有情。

少就是多

Less Is More

“少就是多”是建筑界的一句名言。原来这棵小桃树枝叶甚是繁茂，但枝杈横生无形无态。植物园负责人贝恩德·基希纳先生一直关注着园内植物的生长，经他修剪后形成此景，恰到妙处！

我认为恰当就是好，适度就是美。什么是“度”？就要看每个人的感觉了。此景虽小，但它隐喻着一种观点。试看我国有些名园，虽属“瑰宝”，但园中常常种植着过多、过杂的各种灌木，千姿百态，喧宾夺主。有时会在一些重点地段乱摆盆景，秋季菊展、夏布奇石，极尽堆赘之能事，往往弄巧成拙，大伤雅兴。园林之贵在于情，高于雅，热闹二字是园中大忌。

十多年前曾有幸在泰安参观汉柏院，然而，在千年古木旁竟然配植了繁杂的疏篱、塔柏……林林总总，令人应接不暇。说是陪衬，实属抢景、乱景，把一个国宝级的古树淹没在一文不值的杂绿之中，无宾无主，适得其反。所以，什么是美？什么是度？不仅仅是设计人员的事，更要靠管理者施大恩。北京天坛的祈年殿，旁无杂衬，傲立苍穹，势若“天尊”。上海的方塔园，塔下基座及围墙的处理，状若托盘，恭奉着宋代古塔，尤显得更为神圣。这都体现了恰当及适度构成的美。

为了使这一主题突出，以少胜多的设计理念在“山不在高”中，以及“问道于禅”的景观处理手法均有所体现。

景虽小，
但贵在于情，高于雅。

少就是多

由建筑大师路德维希·密斯·凡·德·罗提出的：「Less is more」。他坚持「少就是多」的建筑设计哲学，在处理手法上主张流动空间的新概念。他的设计作品中各个细部精简到不可精简的绝对境界，不少作品结构几乎完全暴露，但是它们高贵、雅致，使结构本身升华为建筑艺术。「少就是多」，这句话的含义我们也可以轻易地从几千年的中国传统美学和哲学中品味出来。国画大师最有意境的东西往往不是涂满笔墨的画幅，而是在于那一大片空白之中。

Less Is More

"Less is more" was proposed by the architect Ludwig Mies van der Rohe, who insisted on this philosophy in his architecture and in his approach and advocated a new concept of flowing space. In his design work the details had been streamlined to an absolutely irreducible state. Many of the works are almost completely exposed, but they are noble and elegant, which has elevated the design process itself to architectural art. And the meaning of "less is more" can also be easily sensed from thousands of years of Chinese traditional aesthetics and philosophy. The greatest artistic conception of Chinese master painters is often not a frame filled with brush ink, but with large empty spaces.

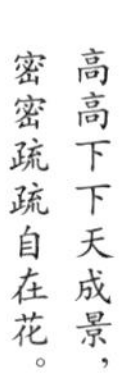

"Less is more" is a saying among garden designers. Originally the leaves of this little peach-tree branch were very lush, but then they became invisible and shapeless. The head of the Ruhr University botanical garden, Mr. Bernd Kirchner, has been continually paying close attention to the growth of the plants in the garden, and since he trimmed it, this scene is just wonderful!

I believe that suitable is splendid, that beauty lies in a good fit. But what is the "degree"? It depends on how everyone feels. While the scenery here is sparse, it has meaning. China has a few well-known gardens which, while they may be gems, just have too much planted, the arrangement of shrubs is too complex, the forms too many, so that [using a Chinese expression] the loud guest upstages the host. Sometimes one will see bonsai, chrysanthemums, and odd, complex arrangements of stones strewn about some important places, and they are superfluous. The value of a garden lies in its emotions and its elegance; the word "bustling" is a taboo in the garden.

More than ten years ago, I was fortunate enough to visit the Han Cypress Tree Courtyard in Taian, a city of Shandong Province. However, there were so many hedges, cypresses and so on, in addition to the millennium-old ancient trees. Said to be a contrast, it is really a scene full of "grabbing" and chaos, submerging a national-treasure-level ancient tree in a worthless sea of green, with neither "guests" nor a "host", for whom only the opposite would be appropriate.

So what is beauty? What is the sense of limit, of degree? It is not only the designer's business; one also needs a manager with a sense of grace. Hall of Prayer for Good Harvest at the Beijing Temple of Heaven [a religious building complex in Beijing] has no miscellaneous lining next to it yet stands proudly in the sky. In the Square Pagoda Park in Shanghai, Fangta Garden, the base of the tower and the surrounding walls are arranged like trays. They pay respect to the ancient towers of the Song Dynasty, and are themselves even more sacred. This reflects the beauty of proper and modest composition.

In order to make this theme stand out, the design concept of "less is more" has been reflected via "the mountain need not be high" and the landscape-treatment method of "seek the way through contemplation".

篱落疏疏一径深，
树头花落未成阴。

独怜幽草涧边生，
上有黄鹂深树鸣。

游廊漫话

Walking along the Gallery, Chatting Freely

沿廊前行渐入佳境，曲折蜿蜒而步移景异。这儿是条石压沿，青砖铺地、木构粉墙、素柱明梁。游廊舍去了多余的装饰，渗透着纯朴明快的民居格调。不同的季节，风姿各异，荫绿灿红，各抒情怀。

八角窗前，原为五角亭（风月亭）立于此。由于园址变动，环顾四周皆无高楼扰眼，小亭则移向右岸水边，有巨石环绕，自成一景。小亭移开留下豁口，并未延续原有座栏封实，仅横一石条，不经意间留点痕迹，有心的游客会产生小小的疑问，游人与景之间就有所互动。豁口与风月亭，似乎隐约着一种无形的联系，景物之间也略显含蓄。

园林设计与戏曲、绘画以及书法艺术均有共通之处，过于完美往往会陷入雕琢。儿时习毛笔字，老人总说：“写字不许描！”这是最初的艺术口诀。书法作品中留些枯笔，有时胜于浓墨；戏曲中偶露沙嗓，会更具个性；国画大家绘的残荷，会比满叶皆绿更富诗意。

世间事物，千变万化，不宜由一定式决其美丑，一切都要相对而言，分别情况辩证对待。罗丹打断了巴尔扎克那栩栩如生的手，反而更加突出了巴尔扎克的形象。不知多少雕塑家为了维纳斯的断臂费尽心机，结果反而是画蛇添足均告失败。缺陷者不等于丑，反具有真实的美，自然之美。乔布斯的苹果被咬掉一口的标志就会使人产生各种版本的猜想，它比完整的苹果更具魅力。缺陷者，美也！

龚自珍的一首诗，透彻地讲述了这个道理：

未济终焉心缥缈，
百事翻从缺陷好。
吟道夕阳山外山，
古今谁免余情绕。

万事不宜皆如意，留有残缺自美丽。欲不可满，乐不可极，莫到琼楼最上层……这些都是常识，也是哲学问题，美学亦同此理。

黄花深巷，红叶低窗，凄凉一片秋声。

星移物换，事易境迁。

Walking along the gallery one gradually enters a wonderful environment, and meanwhile the scenery changes along with the twists and turns. Here are the edges of stones, paving with blue bricks, white wooden walls, and plain beams. The corridor lacks excessive decoration, and its bright yet straightforward stylings permeate the house. Different seasons, different styles, shaded green and bright red, each blends naturally with the environment.

In front of the octagonal window stands the original pentagonal Wujiao Pavilion (also known as the Fengyue Pavilion). Due to the change in the park's location, there are no high-rise buildings around it, and the small pavilion has been moved to the water on the right bank, surrounded by huge stones, creating a scene of its own. Here the pavilion was removed to leave a space, but without retaining the original ring of stones. And there was merely a stone strip, inadvertently leaving a bit of space. Attentive visitors will be motivated to ask questions, and there will be an interaction between them and the scenery. The gap and the Fengyue Pavilion seem to have a vague, intangible connection, and the scenery is slightly veiled.

Garden design has the same features as opera, painting and calligraphy. If it is too beautiful, it tends to fall into the trap of being excessively ornate. As a child, one learns brush writing, yet the elder always says, "When writing, no adornments!" This was the original artistic regimen. One sees some calligraphy works done with dry pens, yet sometimes this is better than using dense ink; occasionally, a raspy voice in a drama will lend more personality; the leftover lotuses painted by great masters in Chinese paintings can be more poetic than scenes full of green leaves.

Things in the world are ever-changing, and it is not proper to determine their beauty or ugliness according to a particular formula. Everything must be considered relative to the circumstances. Rodin broke from Balzac's lifelike hand, but in doing so gave even more prominence to Balzac's image. I don't know how many sculptors exhausted themselves based on Venus' de Milo's broken arm, but they all failed by doing nothing more than gilding the lily. A flawed image of a person does not mean he is ugly. He has real beauty and natural beauty. The logo of Steve Jobs' bitten-into apple will give rise to every kind of conjecture, and is more thought-provoking than a complete apple. To leave something out is beautiful!

A poem by Gong Zizhen [an intellectual of the early 19th century] penetratingly tells this truth:

When you have failed at some purpose and your heart is forlorn,
Remember that it is the best if everything is imperfect in some way.
Chant until the sun sets behind the distant mountains,
From the beginning of time no one has avoided being entangled by regrets.

Nothing is ideal, and one cannot but leave behind an incomplete beauty. Neither desire nor music can be completed, and it is impossible to reach the highest floor of the Qionglou Palace [an imaginary building]. This is common sense, and these are issues of philosophy. Aesthetics is this way.

动摇风景丽，
盖覆庭院深。

别梦依依到谢家，
小廊回合曲阑斜。

屋舍俨然

The Houses are Neatly Arranged

“屋舍俨然”一语取自《桃花源记》。潜园之中游廊曲折起伏穿梭于厅、院、屋舍之间。时而穿过小院，时而悬架水上。景色变幻小中显大，此处是豁然开朗，彼时又略显迷离，透过景窗可以窥见若明若暗的那方景物，半遮半掩徐徐诱人。游廊是连接各处的脉络和景观。

风月亭移至对岸，与主厅隔水而立，彼此呼应，恰为对景。这儿观景，一年四季各显瑰丽，晨夕晴雨不同情趣。此是游人驻足和小憩的最佳处所。由于它们之间相对而立，在潜园中构成全园的视觉中心。

它们可近看游廊粉墙，远眺野渡草棚，四周隔水相望，虽然各具特色，但主次分明错落有序，俨然显现出一幅完整的建筑组群景观。

The words “The houses are neatly arranged” are taken from “Peach Blossom Spring”. The gallery in the garden zigzags through the halls, courtyards and houses. Sometimes it goes through the courtyard, sometimes on top of the water. The scenery changes randomly; small things then large ones, or it is suddenly bright in one place, and then slightly blurry. Through the window, you can see the bright and dark sides of this half-covered, seductive scene. The gallery is the context and landscape that connects all places.

The Fengyue Pavilion was placed on the opposite bank and stands across the water from the main hall, each echoing the other, a scene of two opposites. The views here are magnificent all year round, with different appeal in the sun and the rain, or at dawn and dusk. This is the best place for tourists to stop and rest. As the two structures stand opposite each other, they form the visual center of the whole Chinese garden.

They can take a close look at the gallery’s white wall, overlook the Yedu thatched roof, and face each other across the water. Although each has its own characteristics, the primary and subordinate parts are divided into those scattered about and those that are orderly. Together, a complete landscape suddenly appears.

闲花深院听啼莺，
斜阳如有意，偏向小窗明。

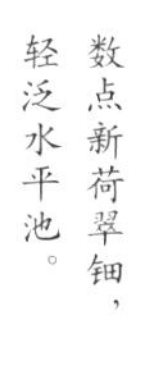

数点新荷翠钿，
轻泛水平池。

庭院深深深几许，
云窗雾阁春迟，
为谁憔悴损芳姿。

上有青冥之高天，
下有渌水之波澜。

秋风秋雨小亭台，对饮举杯尽开怀。

园中一隅

One Garden Corner

60

中国园林宜步步是景、处处入画。在设计时，园中的建筑、景窗、水洞、石块均有所标定。但红透了的秋色是大自然赐予的美。这些都是在园林建成后，植物园主管逐年补充绿化配置而形成的。

红艳艳的画面，让人触景生情。此时，杜牧的“停车坐爱枫林晚，霜叶红于二月花”的千古美句会油然而至。

如果我们把中国园林比誉为“美女”的话，尽管她天生丽质、眉清目秀，但若缺饰少戴，也会是眼大无神，美中不足。植物配植得体方能画龙点睛，这是为中国园林增情添韵的重要环节。

植物的色彩最能表现季节的变化，每当春夏来临，时常会在墙头、廊角处，呈现给游客意想不到的画面。有的景色恬静、有的俊俏，也有的盎然，有的玉叶酥松如淑女般、静悄悄地舒展她们各自的美丽。

Chinese gardens should have scenery at every step, with many paintings. At the time of design, the buildings, windows, bodies of water, and stones were all included. But the beauty of the penetrating red of autumn is nature's gift. All of these were formed after the garden was completed, and the larger botanical garden's supervisors supplement the natural greenery's arrangement each year.

Red and colorful imagery touches people in a particular way. At this point, Du Mu's [a Tang Dynasty poet] ancient, beautiful lines naturally come to mind: "Stopping in my sedan chair in the evening, I sit admiring the maple grove".

If we compare Chinese gardens to a beautiful woman, despite her natural beauty and beautiful, delicate eyes, if she lacks adornment and dresses less elaborately, she will be dispirited. Plants are planted properly to add such a finishing touch. It is an important part of adding charm to Chinese gardens.

The color of a plant can best reflect the changes of season. When spring and summer come, one often sees it on the walls and in the corners, giving visitors an unimaginable picture. Some of the scenery is quiet, some is handsome, and some is magnificent. There are ladylike jade-leaf pines, each quietly stretching out its own respective beauty.

杏子梢头香蕾破。
淡红褪白胭脂涴。

叠石作小山，埋瓮作小潭。

到处皆诗境，
随时有物华。

主厅参差

The Unevenness of the Main Hall

风月亭内看水厅，它与敞厅高低错落，前后相依，组合成潜园的建筑复合体，统称为“主厅”。这种似对称又不对称，虽不对称又均衡的建筑构图，使主体不宜过大得到了恰当的体现。厅内有对联为“清风明月本无价，远山近水皆有情”，描绘出这儿是赏景和聚友的重要场所。

敞厅升高。高架台基的做法，在国内往往是设计师根据个人的感觉信手勾画出石块的组合，施工时建筑工人根据现场石料选择砌筑。然而在德国又是另外一番情景了，基础由德国公司进行施工，他们将这手绘石块组合的图纸，按照 1 : 1 放大，由工厂逐块切割，并一一编号，再运至现场对号入座予以固定。由此也可看出德国人的一丝不苟、严格认真的精神。

但根据我在欧洲各地的参观考察，许许多多出色的石头建筑，无论是级配的组合还是艺术的表达上，都非常精彩。如果我们把话说俗了，就是他们玩石头比我们玩得更漂亮，因为欧洲人在建筑上应用石头已经有几千年的历史了。潜园施工中所以如此做法，我的理解是，无非是因他们对中国园林的谨慎，以及对中国建筑的尊重吧！

Looking at the viewing spot on the water in Fengyue Pavilion, it and the adjacent hall are uneven; each depends on the other to form a building complex in the garden, which is collectively referred to as the main hall. This kind of architectural composition, with symmetry and asymmetry, although balanced, becomes unsuitable if too large. There is a couplet in the hall, “The gentle breeze and the bright moon are priceless in their own right, and the faraway mountain and the nearby water both move us”. This indicates the place is an important one to enjoy the scenery and for friends to gather together.

The open hall is elevated. The designers’ method for elevated platform foundations is often to draw by hand a combination of stones based on the designers’ personal inspiration. During construction, workers choose masonry based on the site’s available stones. However, in Germany, it is a different scene. The foundation was constructed by a German company. They enlarged the stone-combination drawings to the full size, preserving the original proportions, cut the stones one by one in the workshop, numbered each of them, and then transported them to the site to set them. This also illustrates Germans’ strict and meticulous way of working.

But, based on my visits and investigations throughout Europe, I saw many excellent stone buildings, both with respect to the technical level and artistic expression. To put it bluntly, they “play with” stones more beautifully than we do, because Europeans have used stones in architecture for thousands of years. According to my understanding, the reason this method was used during construction was simply their prudence with respect to Chinese gardens and respect for Chinese architecture!

池塘春暖水纹开，堤柳垂丝间野梅。

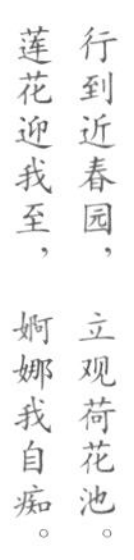

行到近春园，立观荷花池。
莲花迎我至，婀娜我自痴。

绿漪堂前湖水绿，
归来正复有荷花。

春未老，
风细柳斜斜。

砖雕似古

Ancient-Style Brick Carvings

入口前影壁的园洞，原设计是由四块砖雕组合而成，上刻有园内景观图案。园林移此，仍旧复位反而会遮挡后面的美景，但四块砖雕如何安排，放置何处？经现场反复磋商后，顿受启发，即刻联想到曾在意大利一个教堂办公室内拍过的画面。于是我们即刻砍边凿角，并于泥土之中揉擦做旧，随时审视其形，使其状若尘封之“文物”。将它们镶在墙上，犹如博物馆中的展品，诱人遐想。好像历史上曾经存在过这个景物似的！此做法不但丰富了墙面的景观，而且使原来几乎可弃之物，经稍加处理便“身价百倍”。同时，它又为新建的园林增添了历史年轮的沧桑感。

有人会说，此为赝品难登大雅！其实不然，既非拍卖，又何需原物呢？“桃花源”在国内就有几十处，何处属原物？究其原著也只是一种想象。《梁祝》的故事已是家喻户晓，但它有诸多版本，有人考证，梁山伯与祝英台二人竟然是相隔八百多年不同朝代的人，只因一次蹊跷的同地重葬，经文人杜撰成为戏曲，能说它不美么！福尔摩斯是享誉世界的“名人”，谁会追究其为艺术假想呢！

当然，我们的做法也有“度”的问题。但许许多多的照片中，以及在德国出版的《画谈潜园》一书里，这个景，都占据了重要版面。由此可见他们已经认可了这一做法，他们也喜欢上了这一“赝品”组合了。

为了增加游兴，丰富想象，在施工中如此处理应该是值得肯定的。何况在墙上嵌历史装饰品，又是欧洲人的一种时尚，这也可以说是“取之于斯，用之于斯”吧！

The gap in the garden's wall in front of the entrance was originally to be designed by combining four brick carvings and engraving them with landscape patterns. The garden moved them, but they still obstructed the beautiful scenery behind them. How were the four brick carvings to be arranged, and where should they be placed? After repeated on-site discussions, we were reminded of some photos once taken in a church office in Italy. So we immediately cut the edges off and chiseled the corners, rubbed them in the dirt to make them look older, while from time to time randomly inspecting the shape, with all of this done to make each one resemble a relic. Placing them on the wall made them like an exhibit in a museum. As if this scene had really existed in history! This method not only enriched the landscape around the wall, but a slight adjustment also made the original, almost disposable objects "a hundred times more valuable". At the same time, it added a sense of the vicissitudes of history.

Some people would say that this is fake! In fact, it is not something to be auctioned off, so why do you need the original? There are dozens of "Peach Blossom Spring" gardens in China. Where is the original? After all, any study of the original work would just be imaginary. There are many versions of the story of *Liang Zhu*. Some people have verified that Liang Shanbo and Zhu Yingtai were from two different dynasties more than 800 years apart. Because of a reburial, the two dynasties were ahistorically reunited, and a writer wrote a fictional drama. Will you say it's not beautiful? Sherlock Holmes is a world-renowned "celebrity"— who would investigate him merely as an artistic fiction!

Of course, our approach also had a problem of "degree". But from many photos, and from the German version of this book, we can see this scene occupies an important place in its pages. It can be seen that they have approved of this approach, and they also like this "fake" combination.

In order to increase tourism and enrich people's imagination, such management approaches in construction deserve our approval. Besides, it is a European style to inlay historical decorations on walls, and this can be said to be "When in Rome, do as the Romans do"!

土道蚀残砖、
沙草埋折戟。

古拙砖雕、形同文物。

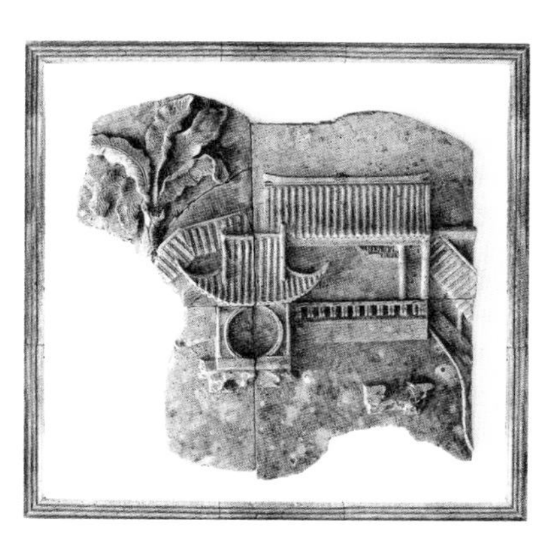

步入幽境

Step into Solitude

12

当我第一眼看到台阶落叶照片这个画面时，真的异常惊喜！即刻联想到《桃花源记》中的“芳草鲜美，落英缤纷”。门外的台阶本属平凡之处，但经大自然的一次点缀，竟然如此神奇、俊美！德国的摄影师并不一定知晓陶渊明的诗文，但他们捕捉的镜头，却和诗中描述的这般契合！可见古今中外对美的感受竟如此相通。

旁观者问：“这几乎不像真的？”但我认为，此绝非是有人“祈福所为”。其疑问本身就告诉我们，它美的程度了。此时是落红片片色彩纷呈，如诗如醉，秀色可餐，奇美也！我想东晋时期的陶渊明所描绘的“桃花源”，也不一定会有如此烂漫吧？

走出主厅，即步入“荒野”，这儿称得上是“苔痕上阶绿，草色入帘青”。如诗一般的恬淡、似画一样的自然。真可谓是“于平凡之中见惊奇”。石阶虽显野趣，但它一石一缝都有讲究。石形的粗细选择，石块大小的级配，似乎有某种说不出的美学规律。在这里，施工人员的感觉和经验是至关重要的了。

从园林布局的角度看，此处是由室内走向室外，由建筑走向自然的过渡地带。石灯、石块、水草、湿地以及无言的花窗，它们都静静地诉说着一切。

When I first saw this picture of fallen leaves on the steps, I was really surprised! I immediately thought of the lines from “Peach Blossom Spring”: “The fragrant grass, fresh and beautiful, fallen petals lie in profusion”.

The steps outside the door were merely ordinary, but once embellished by nature, it was so magical, very beautiful! German photographers don’t necessarily know the poems of Tao Yuanming [who wrote in the time of the Eastern Jin] but the shots they captured with their cameras fit so well, just as described in the poems! We can see that Chinese and foreign feelings about beauty are so interlinked.

Visitors have asked, “Is this not real? It almost seems that way”. But I don’t think anyone is praying to a deity; the question itself tells us how beautiful it is. This time of year is full of autumn red, a time also replete with poetry and beautiful colors! I think even the “Peach Blossom Spring” described by Tao Yuanming in the Eastern Jin Dynasty was not necessarily so brilliant, was it?

When walking out of the main hall, one enters a “wilderness”, which has been described [by Liu Yuxi, a Tang Dynasty poet] as a scene in which “Traces of moss cover the steps green, the green grass enters into nature’s canopy”.

This is a scene as quiet as a poem, as natural as a painting. It is really “seeing surprise in the ordinary”. Although the stone steps express a certain rustic charm, particular attention is paid to each stone. The choice of the thickness and gradation of the stone given its size seem to follow some unspoken aesthetic law. The feeling and experience of the construction crew are crucial here.

From the perspective of the garden layout, here is a transition zone from indoors to outdoors, and from construction to nature. Stones, stone lanterns, water plants, wetlands, and silent lattice windows, they all quietly tell everything.

下有无事人，竟日此幽寻。

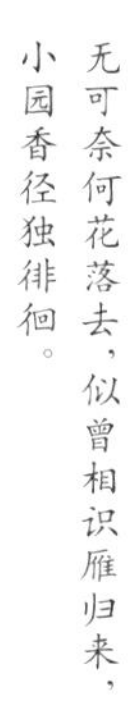

无可奈何花落去，似曾相识雁归来，小园香径独徘徊。

苔痕上阶绿，草色入帘青。

春深黄口传窥树，雨后青苔散点墙。

曲径通幽处，禅房花木深。

即兴小景

Improvised Small Scenes

“采菊东篱下，悠然见南山”是该处景观设计时的初衷。当年的菊花早已逝去，而恬淡、静谧的氛围仍留存在这儿。

葫芦形窗花，在设计图中，尚未明确地绘制。但在施工过程中，当放好井圈接通水源后，感觉井孤墙寡有所欠缺，为此增添了即兴小景，两相呼应，自成一体。多年之后植物的生长姿态生动，春夏秋冬各显美姿，叶茂枝枯均可入画。虽然此为边角小品，但俏丽动人。而今，它已是潜园之中不可或缺的一处景观了。

由此看来，园林不单要靠设计，也需要靠经营。经营得好，方可“点石成金”。

“三分匠，七分主人”是《园冶》中的一句术语，有一种注释，主人是指设计人，但我认为设计人只能算半个主人，中国园林自古即为文人墨客、官宦士大夫之所有，他们不但白天欣赏，也常常是“秉烛夜游”，充分享用。这些人大多精诗文、擅绘画，有能力和情趣在长期享用的过程中反复斟酌，不断推敲，才能日臻完善，逐渐形成享誉中外的园林瑰宝。这是一个长期的过程，设计人焉能包揽全程？此乃为我之切身体会，或许属一孔之见吧！

“Picking chrysanthemums by the eastern fence, I lose myself in the southern hills”. This [from another poem by Tao Yuanming] was the original intention of the landscape design here. The chrysanthemums of those years have long since passed away, yet the tranquil and quiet atmosphere still remains.

The gourd-shaped window grille was not clearly drawn in the design. However, during the construction process, after the well was connected to the water source, I felt that the well’s lonely wall was lacking something. For this reason, a scenery was added on an impromptu basis, and the two echoed each other and from this became one. After many years, the plants’ growth was vivid — whether in spring, summer, autumn, or winter it was beautiful, and leafy branches could enter into this “painting”. Although this was a small work in a corner, it is rather moving. Today, it is an indispensable landscape in the garden.

From this perspective, gardens depend not only on design but also on management. Only good management can turn straw into gold.

“Three parts craftsmen, seven parts owner” is a line from *The Garden Treatise*. But there should be a note: one designer can only be regarded as half of a owner. From ancient times gardens were owned by literati and scholars. Such people not only enjoyed them during the day, but also often liked touring them at night, thus getting the most out of them. Most of these people were well-versed in poetry and painting, and had the ability and interest over the long term to think over and change the way they enjoyed them, in order to reshape, so as to perfect, them. In so doing they gradually established the reputation domestically and abroad of Chinese gardens as gems. This is a long-term process — how can the original designer on his own travel the entire road? This is my personal experience, perhaps it is only a limited insight!

采菊东篱下，
悠然见南山。

似花还似非花，
也无人惜从教坠。

花落月明庭院。
悄无言、魂消肠断。

山际见来烟，
竹中窥落日。

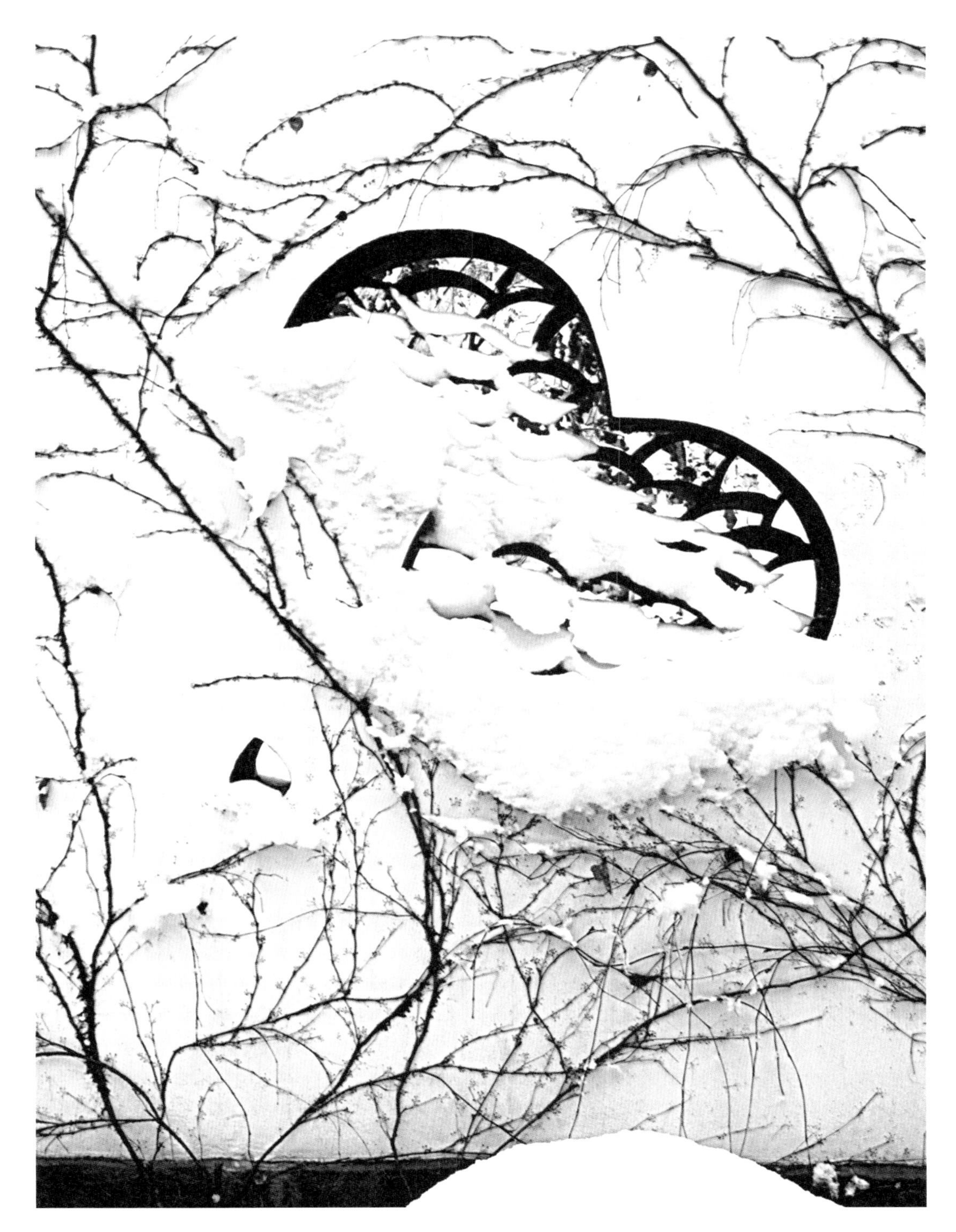

惆怅东栏一株雪，人生看得几清明。

野渡

Yedu (Pulling Oneself Across)

“野渡无人舟自横”是一句脍炙人口的诗句，它出自唐代韦应物的《滁州西涧》。以这句诗命名此处景点，是再恰当不过了！

野渡的地坪，我们在原设计图中是由粗犷的方木组合而成。但是在堆山的过程中，视山势走向由高向低错落转折，蓦然感到生动而自然，势若天成。此时仍再依原图施工，必伤其气势，恐成“蛇足”，权衡轻重，即刻顺势利导，决意延续山形走势，使石矶、石屿如弈布子，点立水中。勾勒出山水相容、曲折参差的空间效果。草棚之下让出泊位，虽无船在，似待舟自来，“此处无声胜有声”的意境不期而至。这是在园林景观施工中，临阵发挥的又一处例证。

硕大的山石，裂隙纵横，卵石入缝，鬼斧神工，观之思绪万千。它很自然地会使人遥想到山洪爆发、沙石俱下的情景。山外有山的意境自然而来，游人立此可意达园外，所谓园林设计中的“小中见大”就包含着想象。造园中的名言“移天缩地在君怀”，不单指的是景，有时更重要的是情，情景交融方属佳构。潜园设计既取自江南园林的“秀”，也试图表现北方山水的“雄”。

“I pull myself across on an unmanned country ferry” is a line from a popular poem. It is from Wei Yingwu’s “Chuzhou West Stream”, written during the Tang Dynasty. It is most appropriate to invoke this poem to refer to this scenic spot!

In the original design, the ground floor of Yedu was composed of rough wooden squares. However, in the process of incorporating the hill [a traditional technique in constructing Chinese gardens], due to its shape and incline, it felt natural, lively, a little evocative of heaven. At this time, the construction was still based on the original drawing, and continuing in this way would slow our momentum, which we feared would render our work useless. Thinking about the trade-offs, we decided to continue to adhere to the shape of the hill, so that the stone breakwaters and islands would be like [Chinese] chess pieces standing in the water. We wanted to think in terms of the spatial effects of compatible landscapes and compatible twists and turns. Under a thatched roof, a berth was left for a boat. Although no boat will ever come, a mood of “silence is better than sound” unexpectedly does. This is another example of lighthearted play in the design of the garden landscape.

In the large rocks, the fissures run vertically and horizontally, there are pebbles in the seams, and the workmanship was superior. This liberated people’s imaginations. One might think of flash floods in the mountains, or of sand sliding in a desert. The artistic conception of other mountains beyond this one will come naturally, and the minds of visitors can easily extend outside the garden. The idea of “seeing the big in the small” in garden design promotes imagination. The famous saying in garden design, “The master moves the sky by shrinking the earth”, not only refers to the scenery; sometimes the most important thing is the emotion. The best arrangements can be found in the spacing and blending of the pieces. The design of the this garden is not only taken from the beauty of the Jiangnan garden, but also from the attempt to express the “grandness” of the northern Chinese landscape.

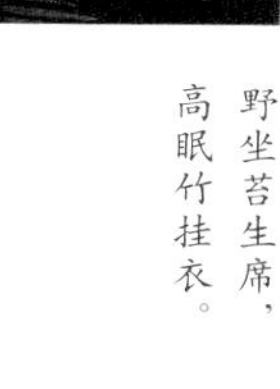
野坐苔生席，
高眠竹挂衣。

野渡无人舟自横

这里有个典故，有个喜欢中国文化的德国老先生，问我「野渡无人舟自横」的出处是王维吗？当时，对于这首诗句我也是不求甚解，顺势应付了一句，王维是写风景诗的。结果，这位老先生回去还真是查出了这句诗的出处，并复印了影印本送给了我。信手拈来的诗句，竟然说不明出处，没有半句语言的教诲却使人永生铭记。后来我们就成了「忘年交」。野渡，这儿用以虚带实的手法，它虽然不同于一般概念中小桥流水，又异于日本园内的「枯山水」，但其「笔触」粗犷，浑厚潇洒的「特写」效果，更符合中国山水诗画之气。这也是在设计和建造时着意追求的意境和想法。

I pull myself across on an unmanned country ferry

There is an allusion to an old German gentleman who likes Chinese culture, and asks me, “Was the person who' pulled himself across on an unmanned country ferry' Wang Wei?” At the time, I did not know much about this verse, so I took the chance to deal with it. Wang Wei wrote landscape poems. As a result, the old gentleman went back to find out the source of the poem, and sent a photocopy of it to me. The poem excaped from my lips without knowing where it came from, but we still could memorize them for life, even though we didn't know who wrote the verse. We eventually became old friends. The theme of crossing the river is a way of using theory to influence practice. Although it is different from the general concept of small bridges and flowing water, it is also different from the “arid landscape” seen in Japanese gardens. But the result of the unrestrained “naturalness” of these “crude brush strokes” is the spirit of both Chinese landscape poetry and Chinese painting. These are also the mood and ideas that are deliberately pursued in both garden design and construction.

野旷天低树，江清月近人。

欲济无舟楫，端居耻圣明。

草棚风雨

Time-Worn Thatched Hut

为什么叫这个题目？有个过程，此处草棚造型在设计时已经决定，但中国工人未做过。鲁尔大学请到远在汉堡的德国专业公司来此施工。尽管我们一再强调要“荒破自然”，但德国的传统做法仍然是顶厚草齐，棱角坚挺，犹如刀切，缺少自然。与“野渡无人舟自横”的意境相距甚远。因此，我们再在顶上敷补草皮，稍予缓解。但时过境迁，十多年的风风雨雨，已将它渐渐地融入这座中国园林中了。

此处景观可以说是中德建筑的结合。草棚是德国做法，根据德国的规范，木柱在室外不能直接落地，草棚的木柱是由镀锌铁件架起，固定在石头上，以防雨水浸蚀而腐烂。

在这里，中德两国的建筑技术和建筑形态，经过时光的洗礼和相互渗透，得到了较好的融汇，竟取得了意想不到的效果。

如今的草棚，霜林黝水略显神秘，风雨晴晦各展英姿，它在潜园之中称得上是独树一帜，远观近瞧各有神态，宅野兼蓄，雅俗适人，它已是潜园内不可替代的独特景观。疏透潇洒的幽野之美，越发会使人联想到“野渡无人舟自横”的诗情画意了。从众多画面可以看出，在不知不觉中，它的美，已征服了来访的游人和挑剔的摄影专家。

Why this title? It was a process. The shape of the hut here had been decided during design, but Chinese workers did not build it. Ruhr University invited German professional companies from as far away as Hamburg to do the construction. Although we had repeatedly emphasized the need here to escape nature, the traditional approach in Germany is to keep the grass thick and firm, as if cutting it were not natural. This is far from the artistic conception of “I pull myself across on an unmanned country ferry”. Therefore, we applied turf to the top and trimmed it slightly. But time has passed, and more than ten years of natural change have resulted in the hut’s gradual integration into this Chinese garden.

The landscape here can be said to be a combination of Chinese and German architecture. The thatched hut is a German practice. According to German regulations, wooden pillars cannot be directly inserted into the ground outdoors. What would be the shack’s wooden posts were erected using galvanized iron parts, which were fixed to the stones to prevent corrosion and erosion from rain.

Here, China and Germany’s architectural technology and buildings’ form, through the baptism of time and mutual interchange, have become better integrated, and have achieved surprising results.

With respect to the thatched hut today, it is unique in this Chinese garden, appearing a bit my sterious when it’s foggy and cloudy, showing different beauty on rainy or sunny days. And when you look at it from a distance, the appearance changes. It is an irreplaceable unique landscape in the garden. The “drunk”, unrestrained beauty of the wild is more and more reminiscent of the poetic and artistic expression of “I pull myself across on an unmanned country ferry”. It can be seen from many pictures that, without knowing it, its beauty has conquered both tourists and discerning photography experts.

花源一曲映茅堂，
清论闲阶坐夕阳。

花隐掖垣暮，啾啾栖鸟过。

败垣芳草、空廊落叶、深砌苍苔。

亦石亦画

Of Rocks and Paintings

建筑造型能表达出一种艺术语言，它会给人以主次远近的不同感受。游人至此，看到的是山石与粉墙的相互交错、建筑与自然的彼此交融。虽然这儿与主厅近在咫尺，但破壁残垣、石破桥寒的表述，屋架石上石立屋中的交错构成，配以荒、破、拙、残等符号的描绘，就能够突显其人迹罕至的意境。例如京剧舞台表演，尽管台上灯火通明，但只要摆上一只烛台或手提一盏灯笼，便会使观众笼罩在夜阑人静、万籁俱寂的气氛之中。舞台如此，园林亦然。

石块与建筑的组合，具体形态在图纸上难于尽表，设计时只能定个大概。如《园冶》中云："峭壁山者，靠壁理也，藉以粉壁为纸，以石为绘也。"但如何画得美妙？难以定式，施工现场的感觉才是实施设计的重要依据。

Architectural shapes can express an artistic language. They will give people different feelings of distance. Here, visitors see interlaced hills and white walls, and the integration of architecture and nature. Although this place is close to the main hall, the expressions of broken walls, broken bridges, "cold" stones, and the staggered structure of the stone towers on the roof slabs, coupled with the depiction of symbols of such things as barrenness, brokenness, clumsiness, and incompleteness, can highlight the artistic conception of its remoteness. Similarly, in a Peking opera stage performance, despite the bright lights on the stage, placing a candlestick or lantern on the set is enough to cause the audience to be enveloped in an atmosphere of midnight stillness, of silence.

The combination of stone blocks and buildings is difficult to express in detail in drawings, and can only be approximated when designing. For example, in *The Garden Treatise* it is written: "On a steep hill, use the wall as support and use the white wall in particular as paint and the stone as paper." But how to paint beautifully? It is difficult to say for certain; the feeling one gets from the construction site is an important basis for implementing the design.

山光悦鸟性，潭影空人心。

日光下彻，影布石上，佁然不动。

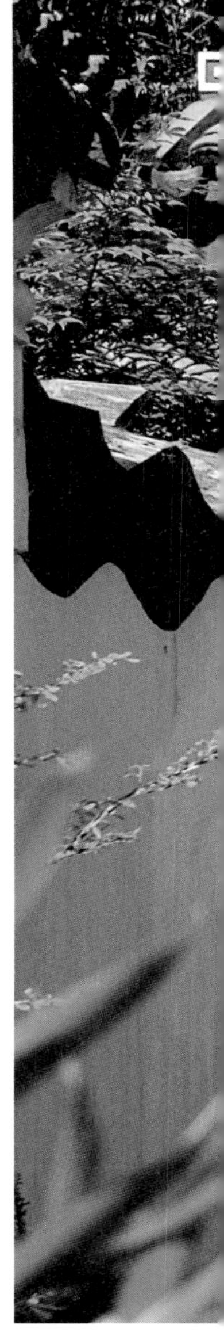

衔山抱水建来精，
多少工夫筑始成。

环保之歌

Song of Environmental Protection

穿过野渡，即来到万石嶙峋的山野地带。在山石之间开涧设洞追求自然，中国园林宜“虽由人作，宛自天开”。在叠山过程中，有个环保的故事值得一书。

德方知中国园林之内需要营造假山，必须准备石料。基建处长福曼先生曾带我们看过几处石矿，但石质、石形都不适合。最后我告知，植物园旁的山石尚可利用。但他们顿时竟异常踌躇。对此我也颇为不解，心想这么简单的事情竟如此为难？！后来我才知道，在德国环境保护法是非常严格的。开采这点山石，在当时的情况下，市长都不能独自决定，为此我们的施工现场竟停顿了数日。后来听说要等议会开会，通过特批，并选隐蔽处，只允许放两炮。唯有经过这一程序，许多石块方能陆陆续续地运至现场 , 为创造雄健幽深的山谷景观，提供了充裕的建造材料。

这是 1990 年的事，为什么德国处处都是水清山绿，一切堪称井井有条，可以说也源于此吧！

After crossing the Yedu, you come to the rocky, rugged area, where gaps in the hills and rocks have been opened to allow one to connect with nature. Chinese gardens should reflect the principle, “Though artificially made, it appears to be nature's work”. In the process of sculpting hills, there is a story of guarding the environment worthy of a book.

The Germans know that rockeries need to be built in Chinese gardens, and stones must be prepared. Mr. Vormann, the director of the infrastructure department, once showed us several stone mines, but the stones' quality and shape were not suitable. I ended up saying that the rocks beside the larger botanical garden were still available. But they suddenly became embarrassed. I was puzzled about this, wondering why such a simple matter was so difficult? Later I learned that environmental protection laws are very strict in Germany. Given this, the mayor could not personally decide to permit the mining of this rock and stone, so our construction stopped for several days. Later, I heard that it was necessary to wait for the parliament to meet to approve the special request, and to choose sites not out in the open. Only two detonations were permitted. Only through this process could many stones be transported to the site, one after another, thus providing abundant building materials to create a strong and deep valley landscape.

This was in 1990, and illustrates why Germany is full of water, mountains and greenery and why everything can be said to be well-ordered; it all emanates from this!

爱此一拳石，
玲珑出自然。

石径行来微有迹，不知满地落松花。

风月亭记

An Account of the Fengyue Pavilion

092 – 093

“风月亭”这个名字是在施工结束时临时起的，因主厅内悬挂对联有“清风明月本无价”的句子，摘取其中两字而取名风月亭。风月亭，原来在游廊的八角景窗前，现移至山脚，两旁巨石相衬，临水而立，若亭内小坐，如在谷间，是小憩的极佳处所。小亭移此，不但使潜园的平面构图更趋完善，而且亦使建筑之间有了呼应感。

在德国园林施工都有严格的规定。中国园林建筑多为传统做法，如按德国标准，这些建筑都不允许建造。因为根据他们的规定，都要经过精确的计算方能施工。

有人问：“为什么亭子做五角形？这是在传统的园林中，很少见的平面。”

回答是：“潜园很小，方形平面与其他建筑重复，圆亭在园内又与多处构图雷同，若六角形显得亭小柱多，而五角亭形态活泼，又无明显的方位感，亦适当丰富了潜园中建筑构图。”

中国园林中，柱子的直径在设计时是值得推敲的，柱子宜细不宜粗。潜园游廊的柱子直径为12厘米，亭的柱子直径为14厘米。柱子细些，屋顶会显得轻巧飘逸，亭子体态更显窈窕，这种效果也符合潜园的主题。

挺拔、俊秀在中国园林建筑造型中，是永远的追求。丹顶鹤形态之美，很大因素是它有着修长而强劲的腿，如果将其变粗反不如鸭了。

造园之道，意在能“品”能“悟”，品才能品出感觉来，悟才能悟出意境来。经过“品”和“悟”的游人更能充分享受到园林艺术的魅力。

连山变幽晦，
绿水函晏温。

绿叶阴浓，遍池亭水阁，偏趁凉多。

The name "Fengyue Pavilion" was adopted for a time after construction. The name "Fengyue Pavilion" was taken because of the couplet sentence "The gentle breeze and the bright moon are priceless in their own right" hanging in the main hall. [The phrase is taken from a poem by the Song Dynasty poet Ouyang Xiu.] Fengyue Pavilion, originally in front of the octagonal window of the gallery, was moved to the foot of the hill, with large stones on both sides, standing next to the water. If the small seat in the pavilion calls to mind being in a small valley, it is also an excellent place to rest. The movement of the pavilion not only made the garden plan even better, but also allowed each building to be a visual echo of the other.

Garden construction in Germany is strictly regulated. Chinese garden buildings are mostly constructed according to traditional practices. For example, according to German standards, these buildings are otherwise not allowed to be constructed because their regulations require accurate calculation before construction.

People have asked, "Why are the pavilions arranged in a pentagon? This is very rare for a planar area in traditional gardens."

The answer is, "The Qianyuan garden is very small, square planes are found in the other buildings, and the round shape is similar to many other elements in the garden. In a hexagon the pavilions would seem small and the columns many. A pentagonal pavilion is lively, and doesn't have such an obvious orientation. It also appropriately enriches the arrangement of the garden's buildings."

In Chinese gardens, the diameter of the pillars is worth considering during design. The pillars should be quite thin, or quite thick. The diameter of the pillars on the gallery in the garden is 12 cm, and the diameter of the pillars in the pavilion is 14 cm. The thinner pillars of the roof appear light and elegant, and the pavilion's "posture" is slender. This effect also conforms to the garden's theme.

In Chinese garden architecture the themes of "tall and straight" and "elegant beauty" are always pursued. The beauty of the form of a red-crowned crane is largely due to its long, strong legs; if the legs were thickened, it would be inferior even to a duck.

The way of gardening is the ability to "savor" and "comprehend". Savoring is required to enable one's feelings to emerge. Visitors are able to "savor" and "comprehend" or more able to fully enjoy the charm of garden art.

满园深浅色，照在绿波中。

石水之间

Between Rocks and Water

陈从周先生在其著述的《说园》一书中云："表面观之似水石相对，实则水必赖石以变。无石则水无形、无态……水本无形因岸成之，平直也好，曲折也好，水口堤岸皆构成水面形态之重要手法。至于水柔水刚，水止水流，亦皆受堤岸以左右之。石固有刚柔美丑之别，而水亦有奔放宛转之致，是皆因石而起变化。"这些都阐述了水对于石、水对于岸以及水对于堤的依赖关系。

但有时它们之间又存在相互依托的辩证关系，所谓"水遇山而健，山因水而活"，就是这种意思，世间之事千变万化，均不宜生搬硬套。造园有"法"而无"定式"，"外师造化，中得心源"。只有深厚的积淀才能举一反三，方可根据情况，挥洒自如。

园林中的水，有流、有汇，也要有源。有动、有静，又要有不尽之意。石有形，崖求势，选形造势，参差自生。山涧宜堆砌出自然之神韵，跌宕险峻之气势，又要有惊涛裂岸，涛过痕出的时间感和动态效果。中国造园讲求自然，讲究气势，更要讲究有含蓄之美。

假山上的瀑布是潜园的进水口，它的回水管则隐藏在影壁前的石头下面，它们会永远保持着潜园水位的恒定。

Mr. Chen Congzhou wrote in his book *On Chinese Gardens*, "The surface looks like water and stone opposing each other, but in reality water depends on stone to be distinct. Without stone, water is invisible and without form...Either straight or crooked, the inlets and embankments are both an important method of shaping the water's surface. And water is also affected by the embankment, becoming 'gentle' or 'firm', flowing or still. The inherent hardness or softness and beauty or ugliness of the stones differs, and yet the water in some sense flows freely, even as it is changed by the stone." These things all explain the dependence of water on stones, shores, and banks.

But sometimes there is a mutually dependent relationship among them, as the saying "Water meets the mountain and is shaped, and the mountain lives because of the water" indicates. The world is ever-changing, and so one should not copy indiscriminately. Gardening has "laws" but no "fixed form"; "It is the external form that creates, in moderation lies the origin of the heart's satisfaction". Only through gradual accumulation can one learn by example, and only then can one experiment freely according to the circumstances.

The water in the garden flows, it converges, and it must have a source. It moves yet is tranquil, and it must be without a terminus. The stones are tangible, the precipice seeks force, one uses the shape to build the reality, and life is created from the differences. A mountain stream should collect nature's spiritual rhythms, and dissipate and moderate harshness. So there must be a sharp break on the shoreline, providing a sense of time's changes. Chinese gardens speak to nature, to grandeur, and especially to subtle beauty.

The waterfall on the rockery is the water inlet for the garden, and its return flow is hidden under the stone in front of the screen wall. Together, this arrangement will always keep the garden's water level constant.

漠漠轻寒上小楼。晓阴无赖似穷秋。淡烟流水画屏幽。

行行叠石正新秋，
山气清凉老火收。

叠石疏泉不数旬，水芝开出似车轮。

庭下如积水空明，水中藻、荇交横，盖竹柏影。

胜日寻芳泗水滨，无边光景一时新。

山不在高

A Mountain Need Not Be Tall

20

“山不在高，有仙则名。”这儿的山虽不高，也更没有仙人，但我们在设计中，可以借鉴古人的经验和实践，丰富和增大其艺术想象。

为了显示山的高，采用了压低围墙的手法，以反衬出山的雄伟。如进入北京天坛的圜丘，当人们走过第一道围墙时，墙略高过肩膀，而走到第二道围墙时，墙顶就是齐胸高了，人行其间，稍加注意就好像会产生逐渐升高的感觉。只有有心人才能品味出其中的奥妙。看一幅画、听一曲琴，每个人都会有高低雅俗的不同感受。园林设计中利用反衬手法表达，也可以使人产生微妙的艺术联想。如果以吃西瓜的速度咀嚼橄榄，那就什么味道都品尝不出来了。

“The presence of a deity makes a mountain, and not its height”. Although the mountains here are not high and there are no deities, we can use the experience and practice of ancient people to enrich and enlarge our artistic imagination.

In order to better display the height of the hill, the wall was graded to reflect the majesty of the hill's height. This is like when one enters the divine hill of the Temple of Heaven in Beijing. When people walk through the first fence, the wall is slightly higher than their shoulders, and when they reach the second fence, the top of the wall is as high as the chest. While walking, if one is paying attention one has the feeling of gradually ascending. Only those who deliberately look for mystery can taste it. Looking at a picture or listening to a piano, everyone will have different feelings over whether it is elegant or vulgar. The use of contrasting modes of expression in garden design can also create subtle artistic associations. If you chew olives at the speed of eating watermelon, you can't taste anything.

萧萧远树流林外，
一半秋山带夕阳。

文化乐园

A Cultural Paradise

21

埃尔马·魏勒教授在为《潜园：波鸿鲁尔大学植物园的中国园林》一书所作的序言中指出：“在这座给学者和学生以及游赏者提供安静环境和丰富内容的植物园内，又增添了一处珍宝，一处当你不断来访时能感受到它那独特而珍奇的园林——潜园。”

而今，令我们设计者感到格外高兴的是，现在的潜园不但是周边城市居民经常造访的旅游点，而且也是青年学生休闲散步和促膝交谈的理想环境，还是一处举办学术活动和艺术聚会的“香格里拉”。

Prof. Elmar Weiler stated in his foreword to the book *Qianyuan: the Chinese Garden in the Botanical Garden of Rur University in Bochum*: “In this botanical garden, which provides a quiet environment and a wealth of separate elements for scholars, students and visitors, another treasure has been added, a unique and rare Chinese garden whose distinctiveness and rareness you can feel whenever you visit.”

Today, we designers are extremely pleased that the current Chinese garden is not only a tourist spot often visited by residents in the surrounding cities, but also an ideal environment for young students to take a leisurely walk and talk intimately. It is also a "Shangri-La" for academic activities and art gatherings.

学术交流

艺术聚会

中国茶室

A Chinese Tea House

22

潜园建在德国，并非政府资金投入，也不是老板开发的营利项目。它是当地的一项公益事业，和植物园一样，每天免费定时对外开放。波鸿市储蓄银行文化基金会出面筹资并参与策划。首期工程是园林主体，已于1990年底建成，二期工程原是学术活动中心，后来德方又将其改为中国茶室。

1993年，鲁尔大学拟策划第二期工程中国茶室。我为其构思了初步草图。二期基地选择建在邻贴潜园旁边的高台之上，使其与第一期的潜园形成略有高低错落、稍有变化的组群构图。潜园通过爬山廊与茶室相接，茶室又有独立的对外大门。二期不单要考虑地质情况，更要考虑到交通条件、能源供应和冬季采暖等诸多问题。但由于多方面的原因，二期工程项目搁置到近期。

2000年波鸿市和鲁尔大学成立了“中国园林协会”，由此对潜园的保护、管理及建设，有了专门负责机构。中国茶室，不但为游人提供了室内休闲、品茶的环境，更是一处优美的文化交流和文艺演出的温馨场所。

The park was built in Germany, not through government funding, nor as a for-profit project developed by the owner. It was a local public-welfare undertaking. Like the botanical garden, it is open to the public for free every day. Its funds were raised by the cultural foundation of the local bank, which participated in the planning. The first phase of the project was the main body of the garden and was completed by the end of 1990. The second phase originally involved the construction of an academic-activity center. Later the German side changed it into a Chinese tea house.

In 1993, Ruhr University was planning the second phase of the Chinese tea house. I conceived a preliminary sketch for it. For the base, it was decided to build on a high platform next to the garden, so that based on the design of the first phase a pattern of rising and falling heights was introduced; this slightly changed the arrangement of the components. The garden was connected to the tea house through an ascending gallery, and the tea house has an independent door to the outside. During this second phase we had to consider not only geological conditions, but also many issues such as traffic, energy supply and winter heating. However, for various reasons, the second phase of the project was put on hold until recently.

In 2000, the “Chinese Garden Association” was established at Ruhr University in Bochum. There were specialized agencies responsible for the protection, management, and construction of the garden. The establishment of the Chinese tea house not only provides visitors with an indoor environment for tea-tasting and leisure, but also a nice place for wonderful cultural exchanges and performances.

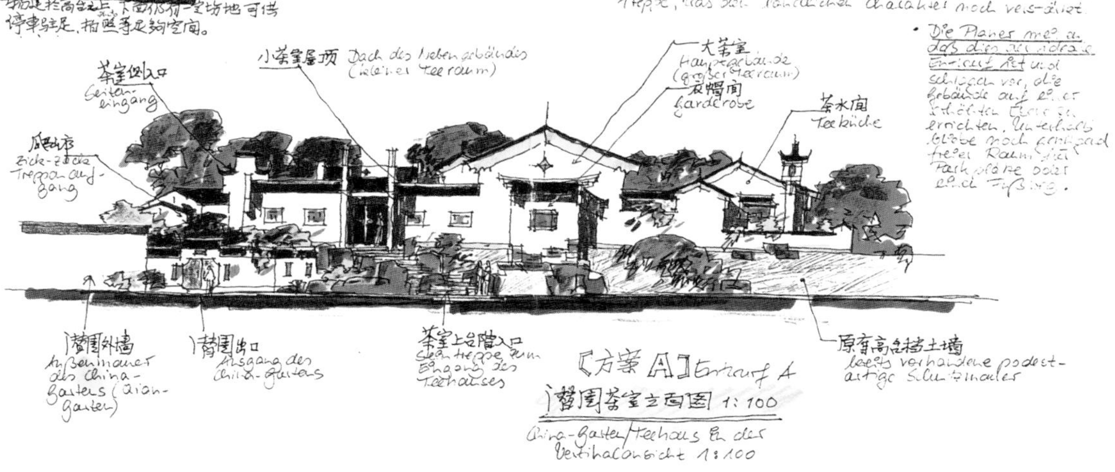

Übersetzung: Susanne Baumann 4/93

茶室初步立面草图

外墙景观

Exterior Landscapes

23

108 – 109

中国园林虽属瑰宝，但过去都属私家所有，从理念和实际情况看，都会有一定的局限性。譬如古典园林的外墙就很少有条件考虑景观，若此时此地造园完全一味仿古，势必欠妥。潜园的外墙，依其主次均作了景观构思。

这是即将完工时的一张照片，矗立水中的影壁与入口建筑方圆相济，彼此呼应，它们与水池、院墙相互组合，构成具有藏风纳气、三向围合的风水前院。半封闭的建筑空间，隔绝着园内的神秘，白色的园墙，永远是流动着树影的画卷。

古诗“林梢一抹青如画”的意境进入了我们的镜头，借诗中那既有前观又有背景的优美构图，来描述我们的这个情景，也可谓林梢一抹“村”如画。

另外，进厅的后壁、沿路的外墙，以及主厅背后的折廊，破墙而出又蜿蜒复入，既传统又破例的建筑处理手法，它们都为外墙景观增添了美妙的画面。

Although Chinese gardens are treasures, they were privately owned in the past. From the perspective of ideal versus actual conditions, they all have certain limitations. For example, the exterior wall of a classical garden is rarely able to consider the landscape. But the exterior wall of this garden in particular was conceived according to the landscape's primary and secondary features.

This is a photo from when it was about to be completed. The screen wall standing in the water and the entrance structure complement and echo each other. Combined with the pool and the courtyard wall, the Fengshui of the forecourt includes a three-directional enclosure, and a three-directional enclosed forecourt with “hidden” received air. The semi-closed space in the structure isolates the “mystery” inside the garden, and with the white will always give one a “flowing” image of the trees.

There is a line from the classical-style poem “The tips of the branches are green as if in a painting” [from the Song poet Qin Guan] that entered our artistic conception. We borrowed from this line its combination of forward-looking vision and background to describe the beautiful scene we composed. This scene can also be described as a “village” of tree tips evocative of a painting.

Additionally, the back wall of the entrance hall, the outer wall along the road, and the winding gallery behind the main hall emerge out of the wall and keep meandering. They are all traditional and exceptional architectural treatments, and they all add a wonderful effect to the exterior wall landscape.

林梢一抹青如画，
应是淮流转处山。

风影轻飞。
花发瑶林春未知。

幽墙几多花。
落红成暮霞。

拂墙花影动，
疑是玉人来。

面壁看红影，
蒲团对绿阴。

四面青山是四邻，烟霞成伴草成茵。

瓦的故事

The Story of the Tiles

24

当我刚看到如此众多的屋顶照片时，有些诧异！司空见惯的屋顶，竟吸引了众多好奇的镜头？

记得 2001 年园林屋顶进行翻修期间，我们准备的瓦片堆积如山。一日下午有几位德国老太在植物园中散步，她们的衣着甚是讲究，其中有两位非常愉快地走近我们，并很有礼貌地表示想要一两片瓦。当得到我们的允许时，她们高兴的心情竟溢于言表。我想，她们是当地居民？还是中国园林的爱好者？或许是园林建设的资助人？不管怎样，这两片瓦一定会成为她们家的摆设。

园林的屋顶上有着数不清的瓦片，形成有韵律的排列，节奏明快，纵横交错美似琴键，屋檐翘角如舞如翩，因此就不难理解，为何有如此众多的生动画面会成为德国摄影师镜头的“猎物”。

瓦片虽小但有时会影响巨大，为何屋顶要翻修呢？

潜园的建造在 1990 年 5 月开工，同年 11 月底竣工，只逾半年。而屋顶铺瓦是最后一道工序，但时已入隆冬，德国气候大异于江南，白天尚暖，一到夜晚气温骤降，砂浆即刻结冰，初始黏结强度极受影响。风侵雪融年复一年，松动、位移、脱落并逐渐漏水，以致多处破损，只得关闭，几乎频临拆除。因此十年之后进行了一次屋顶彻底翻修，此次施工避开了寒冷季节，并增加了防水层，至今十年有多，只在个别处瓦片有所脱位，尚无大碍，择日调整即可。

境外造园，屋顶铺瓦为薄弱环节。国内园林也要每 3~5 年进行屋顶维修，堪称小事；而在德国，他们人高体重，对此手工异常生疏，脚踩屋面如履薄冰。德国的施工，不同工种操作分工极为严格，遇此情况他们会望而生畏，犹如第一个吃螃蟹的人，不知从何下手。此也是国内园林界应进一步研究的课题。

映水曲、翠瓦朱檐，垂杨里、乍见津亭。

旧酒投，新醅泼，
老瓦盆边笑呵呵，
共山僧野叟闲吟和。

When I first saw so many photos of the roof, I was somewhat surprised! Such an ordinary roof attracted so many curious shots?

I remember that during the renovation of the garden roof in 2001, the tiles we had prepared were piled as high as a mountain. A few German old ladies were walking in the botanical garden one afternoon, and their clothes were very proper. Two of them approached us gleefully and expressed politely that they would like a tile or two. When we gave our permission, their joy was effusive. I thought, are they local residents? Or are they lovers of Chinese gardens? Perhaps patrons of garden construction? No matter what, these two tiles would definitely become furnishings for their homes.

There are countless tiles on the roof of the garden, forming (to use musical language) a rhythmic arrangement with a cadence, beautifully crisscrossing the keys of a piano, and the eaves of the roof resembling a dance. So many vivid pictures were available as the "prey" of the German photographer's lens, which is not difficult to understand in a foreign land.

Although the tiles were small, they could sometimes have a huge impact. Why then should the roof have been renovated?

The construction of the garden started in May 1990 and was completed at the end of November of the same year, taking just over half a year. The laying of the roof tiles was the last step, but it was already the middle of winter. The German climate is very different from that of Jiangnan. It is warm during the day. But when the temperature drops suddenly at night, the mortar immediately freezes and the strength of the initial bond was really affected. Every year the wind would blow and the snow would melt and leak in, causing tiles to loosen, and tear or fall off. As a result, the roof was damaged in many places and the garden had to be closed. Because of this, 10 years later the roof was thoroughly renovated. This time work was avoided during the winter and a waterproof layer was added. Since then more than 10 years have passed. In only a few places have tiles been dislocated, and only partially at that. Only minor repairs have been needed.

For overseas gardens, roof tiles are a weak link. Gardens in China also need to perform roof maintenance every 3 to 5 years, which is a trivial matter. But in Germany, the workers are taller and heavier, they are not very familiar with this kind of work, and when walking on the roof, it is like walking on ice for them. And in German construction, the division of tasks among different types of workers is extremely strict. Given all this, they can be intimidated, and no one wants to go first. [The Chinese idiom here is, "Who wants to be the first to eat one of the crabs?"] This is something to be looked into by China's gardening industry.

柳烟丝一把，
暝色笼鸳瓦。

碧瓦偏光日，
红帘不受尘。

一夕轻雷落万丝，
霁光浮瓦碧参差。

妙在因借 Borrowed Wonder

25

118 – 119

这张照片，看了感到神清气爽，不知怎的竟联想到“月落乌啼霜满天”的诗句。鼓足勇气，也即兴诌诗一首（借景）：

雪落潜园瓦垄清，
水岸楼台各西东。
一亩小院无边大？
只缘外林入园中。

诗入不入流？是否规范？暂且莫论，主要是想借此赞颂景色的旷怡之美，并且阐述“借景”的真实效果。

“妙在因借”或“贵在因借”是园林设计中的术语，它们和“嘉则收之”属同类意思。“借景”出自《园冶》一书中的一个章节，是对世界园林史上的一大贡献。《园冶》一书出于明代，但此前在先人的诗歌中都有借景的含意和雏形，如陶渊明的“采菊东篱下，悠然见南山”以及李白的“举杯邀明月，对影成三人”等都有着借景的因素。《园冶》的作者计成，总结和归纳了前人的诗歌和造园实践，形成设计理论列入书中，利在千秋，更应称其为“妙”。

无锡的寄畅园内，可以望见锡山上的宝塔。北京西郊玉泉山的山形塔影，进入了颐和园的游人视线之中。这些都是在园林设计中借景的范例。潜园周围虽没有具体的建筑形象引入园内，但其园外的树冠之高低起伏，疏密有致，绿肥红瘦，各显姿态。借景，有时不但是扩大了空间，更是给游人增加了情趣。在这里，它比具体建筑形象似乎还多了一层“气氛”和“情调”。

从下面的几张图中就可以体会到借景的个中奥妙。

有的会让人联想到“长亭外，古道边，芳草碧连天……”这是最美的歌词了。这也是一幅古时候村头乡外典型的抒情画面。

有的是林木茂盛郁郁葱葱，如在深山老林中。也有的画面若卉木轩窗，翡树琼枝，别有洞天。

据说，拍照、唱歌，都会上瘾，我也有所感触。我这个“弄斧诗人”前面作了七言，现在还想用五言来描述第120页的这张借景图片：

绿柳绘晚风，秋叶别样红，
人约黄昏后，归雀戏林中。

悠哉，悠哉！陶然，陶然！诗虽不才，但我确过了把瘾！

日脚淡光红洒洒，薄霜不销桂枝下。
依稀和气排冬严，已就长日辞长夜。

因借

巧于因借是园林设计的手法。「虽由人作，宛自天开。巧于因借，精在体宜。」这是《园冶》一书中最为精辟的论断，亦是我国传统的造园原则和手段。「因」是讲园内，即如何利用园址的条件加以改造加工。《园冶》说：「因者，随基势高下，体形之端正，碍木删桠，泉流石注，互相借资；宜亭斯亭，宜榭斯榭，不妨偏径，顿置婉转，斯谓『精而合宜』者也」。而「借」则是指园内外的联系。《园冶》特别强调「借景」「为园林之最者」。「借者，园虽别内外，得景则无拘远近」，它的原则是「极目所至，俗则屏之，嘉则收之」，方法是布置适当的眺望点，使视线越出园垣，使园之景尽收眼底。如遇晴山耸翠的秀丽景色，古寺凌空的胜景，绿油油的田野之趣，都可通过借景的手法收入园中，为我所用。这样，造园者巧妙地因势布局，随机因借，就能做到得体合宜。

On borrowing

The rationale for borrowing: borrowing from what is coincidentally there is a technique of garden design. "Though artificially made, it appears to be nature's work, it happens to be borrowed, and its excellence lies in its perfect fit." This is the most incisive assertion in *The Garden Treatise* and it also runs through our country's traditional gardening principles and methods. *The Garden Treatise* also says: "This allows that as the ground rises and falls, the shapes can be corrected, obstructive trees can be removed, water can flow properly through the stones, the components can support one another, and the artistic pleasure in each part is as it should be; the branches lean into the walkway as they should, the objects are placed in an agreeable way, in short everything is suitable and splendid." And so "borrowing" refers to the connections between inside and outside the park. *The Garden Treatise* particularly emphasizes "borrowing scenery" as "the ultimate in gardening", writing further that "If we borrow, even though the inside and outside of the garden are different, the scenery is not restrained by distance." Its principle is "separating the vulgar and absorbing the elegant". The method here is to arrange overlooks inside the garden appropriately to extend the line of sight outside the park. For example, if you encounter something interesting in the beautiful scenery of the Qingshan Mountains, the towering space of an ancient temple, or some green fields, you can extend it into the garden for one's own use, and indeed I have done this. In this way, the garden designer can cleverly arrange the layout according to the situation and borrow as the opportunity arises, and thus realize something that is a perfect fit.

闲寻诗册应多味，得意鱼鸟来相亲。

Looking at this photo, I felt refreshed and somehow thought of the poetic verse “Moon's down, crows cry and the frost fills all the sky” [by Zhang Ji, a Tang Dynasty poet]. I mustered my courage and improvised a poem with “borrowed scenery”:

Snow falls on the tiles of the garden,
On the bank several terraces run east and west.
Can a mu [Chinese acre] of small courtyard be infinitely large?
The only trees in the garden come from outside.

Does the poem flow well? Does it conform to the rules of poetry? For now, I will accept that this is so and think of borrowing this beautiful scenery and describing its true effects.

“Borrowed wonder” and “borrowed nobility” are terms in garden design. They have the same meaning as “separating the vulgar and absorbing the elegant”. “Borrowed Scenery” is a chapter in *The Garden Treatise*, and it is a great contribution to the history of world gardening. *The Garden Treatise* originated in the Ming Dynasty, but its language talked of previous examples of borrowing scenery and prototypes, even from literature, for example Tao Yuanming's “Picking chrysanthemums by the eastern fence, I lose myself in the southern hills”, and Li Bai's [Tang poet] “I raise my chalice to invite the shining moon, the moon casts a shadow, and we have a triad”. The author of *The Garden Treatise* studied, summarized and broadened his predecessors' poems and gardening practices, and turned his design theory into this book. The labor was done then, but its achievement will last forever, and this should be termed “wonderful”.

In the Jichang Park in Wuxi, you can see the pagoda on Xishan Mountain. The mountain-shaped tower of Yuquan Mountain in the western suburbs of Beijing can be seen by tourists in the Summer Palace. These are all examples of borrowing scenery in garden design. Although there are no concrete architectural images that can be seen inside the park, the height of the trees outside it collectively undulates, their density varies, and they are green, round, thick and thin, each having its own shape. Borrowing scenes sometimes not only expands the space, but also adds interest for tourists. Here, compared to concrete architectural design it seems to have an additional layer of “atmosphere”, of mood.

From the following pictures, you can experience the mystery of borrowing from other places.

Some will make people think of the 1935 song lyric, “Outside the pavilion, along the ancient road, the fragrant, jade-like grass reaches the stars... ” This is such a beautiful line, and also a typical scene seen at the edge of villages in ancient times, profoundly emotional.

Some are lushly treed, e.g. in old-growth forests deep in the mountains. There are also scenes such as “windows” of “vegetation”, “trees” of jade, and other beautiful but hidden spots.

They say that taking pictures and singing are both addictive, and I personally am also moved by things made available to me by my senses. I clumsily placed a seven character poem above, and now I want to use a five character poem to describe this picture (P120) with “borrowed scenery”:

The green willow paints the evening breeze, the autumn leaves are especially red,
After one meets the dusk, the returning birds frolic in the forest.

So playful, so joyful! Although my poetry is poor, I really am addicted to writing it!

心无物欲，即是秋空霁海；
坐有琴书，便成石室丹丘。

水绘春秋

Water Paintings, Spring and Autumn

26

倒影乎?

油画乎?

九寨乎?

何美乎?

尽在不言中。

中国园林中的景观是可以设计的，唯有水中倒影难于操控。摄影师的镜头凝固了许多极妙的瞬间，有时竟会交映出酷似印象派的画作，它称得上是对园林之美的一次再创作。

江苏如皋的水绘园，自古就是以水为贵、倒影为佳，故而得名。如今德国的摄影师，将潜园的黄黄绿绿映绘水中——水绘春秋。

Reflections?

Oil paintings?

Jiuzhai [valley]?

What sort of beauty?

The answer to this cannot be verbalized.

Landscapes in Chinese gardens can be designed, although reflections in the water are the only thing difficult to manipulate. The photographer's lens captures many wonderful moments, and sometimes it produces scenes that resemble impressionist paintings. It can be regarded as a re-creation of the beauty of the garden.

The water painting garden in Rugao, Jiangsu, since ancient times has been known for and treasured because of its water; its reflected images are extraordinary. Today, German photographers paint the greens and yellows reflected in the waters of their Chinese garden — water paintings , spring and autumn.

金翠楼台，倒影芙蓉沼。
杨柳垂垂风袅袅。
嫩荷无数青钿小。

山绕平湖波撼城，
湖光倒影浸山青，
水晶楼下欲三更。

天水碧，染就一江秋色。

水真绿净不可唾，鱼若空行无所依。

绿树阴浓夏日长，楼台倒影入池塘。

姹紫嫣红

Beautiful Flowers of All Sorts

27

秋天是丰收的时刻，也是斑斓的季节，红橙黄绿越显魅力，墙上地下各呈缤纷。绚丽、俊美、高贵、鲜艳。把一个小小的园林装点得姿态各异，处处迷人。

赞誉金秋，千言万语难于尽表，姹紫嫣红也不为过，还是借用李清照的诗句吧！

这次第，怎一个“美”字了得！

Autumn is the harvest season and very colorful. The reds, oranges, yellows and greens are more attractive, and both on the walls and on the ground it is beautifully colorful, precious and bright. Decorate a small garden with many different themes, and every location will enchant people.

In praising autumn's gold, there are no words that are adequate! Even "gloriously beautiful" is not excessive! We ought to borrow a line from a poem of Li Qingzhao [Song Dynasty writer of poetry and other works]!

Here, what word for "beautiful" could be adequate?

山明水净夜来霜，
数树深红出浅黄。

树头媚日惯能留，
霜叶轻黄未折秋。

鲜鲜黄叶略如春，
银杏千株晚置身。

一枝挺挺几枝横，
佳客相怜太瘦生。

银装素裹

Silver Adornments, Plainly Wrapped

130 – 131

“忽如一夜春风来，千树万树梨花开。”大雪降在潜园内，点染出一种洁白的美、对比的美、概括的美。远隔万里的德国，受大西洋气候的影响，每年冬季常是大雪纷飞，它会给中国园林增添在江南园林中难得一见的异国情调——银装素裹。

对雪景描述得既形象又别致的诗，当属唐朝的张打油。借此我们将他咏雪的两首诗展录如下:

（一）

江上一笼统，井上黑窟窿。

黄狗身上白，白狗身上肿。

（二）

六出飘飘降九霄，街前街后尽琼瑶。

有朝一日天晴了，使扫帚的使扫帚，使锹的使锹。

生动、俏皮，视角独特，朗朗上口但又回味无穷。难怪打油诗会流传千古。

一场雪后，原来五颜六色的景致，都呈现出了异常含蓄的黑白两色，有人说墨生五色，这儿我们可以说，墨胜五彩。为什么许多摄影家的作品喜欢用黑白照片来展示呢？道理大概就在于此。潜园的众多雪景画面，有时确实会比原物更显得别致和稀有吧！

江上一笼统，井上黑窟窿。
黄狗身上白，白狗身上肿。

六出飘飘降九霄，
街前街后尽琼瑶。

画堂晨起，来报雪花坠。

“Suddenly, the spring breeze arrived, and thousands of pear trees blossomed.” Heavy snows fall in the Qianyuan garden, dyeing it with white, creating a contrasting, and comprehensive, beauty. In distant Germany, affected by the Atlantic climate, winters are often snowy. This adds a rare, exotic, silver-covered atmosphere to a Chinese garden that is difficult to achieve in a garden in Jiangnan.

Poems that described snow in a way both unconventional and full of striking imagery are represented by Zhang Dayou of the Tang Dynasty. Here are two of his poems about snow :

1.
After the snow falls, the sky and the ground alike are white.
And the well seems to be a hole in the snow,
The yellow dog is white, and the white dog seems to grow in size.

2.
The six-petaled snowflakes have fallen from the sky, and up and down the street, all is like jade.
When a clear day comes, we will then dig out and sweep up.

These two poems are lively, playful, and have a distinct perspective. They are catchy yet have an eternal beauty. No wonder simple poems like these are transmitted down through the ages.

After a snowfall, the original colorful scenery presents unusually subtle black and white colors. Some people say that the ink of nature comes in five colors. Here we can say that the ink of nature wins through its colors. [Note: Here the author is making a pun with Chinese characters.] Why do many photographers like to shoot in black and white? This is probably the reason. The many snow scenes shot in this garden sometimes seem more distinctive, rarer than the original!

白雪却嫌春色晚，故穿庭树作飞花。

问道于禅

Asking about the Tao and Zen (Chan)

62

136 – 137

有人说王维的诗有禅意。也有人说八大山人的画有禅意。如何理解“禅”？我也只能算是道听途说。但这张入口前的雪景照片，我确实感到有点儿“禅”的味道。禅宗兴于我国已有一千多年的历史，后来又与儒、道融汇，博大精深。参禅论道，不敢涉足，更不可妄言。但在园林设计中，作为一种意境追求还是应该值得探讨的。

是否可以说，禅是一种美，也是一种智慧。静静空空囊括四海，冥冥灭灭气盖八荒。这里虽“无”，但可胜“有”，少就是多的美学观念也得到了充分体现。

任何事情都应一分为二，禅学再深奥也不可能“悬空”永世，一旦“落地”也会成为被市场瞄准的一种“宠物”。“茶禅一味”的四个大字，已是茶馆酒肆的高雅点缀；“禅道之家”也成了瑜伽会所的金字招牌。可见其民俗化和时髦化到何种地步了！

我为何对这张图情有独钟呢？就是这一氛围是我们设计之初的期望。这里的构思和处理有别于园内。长长的直线，素雅的高墙，黝黑的水池，烘托出点石一二，任人想象，思绪无边。而皑皑白雪又更加概括和提炼了此一构思的艺术效果，故而歌之！

一位智者在信中说：“野渡中的草棚之下让出泊位，虽无船在似舟自来的做法，就具有‘禅’的意境。”“此虽无一人，也无一舟，更无一桨，但它却蕴含着独有的具象美。以空显有高于有，以虚代实美于实。”

冥冥之中若有若无，含蓄的表述胜于直白，“禅”更是一种境界。

诗境何人到，禅心又过诗。

禅

一般很少有人会写「禅」，而我写「禅」有两个原因：第一，设计过程中有「禅」的味道；第二，德国人认为禅宗是日本的，因为日本有翻译禅宗的书到欧洲，故此，欧洲人认为日本是禅宗的起源。日本的「枯山水」是很成功的，是禅的味道，禅的延续，但其实「禅宗」是在宋代由中国传入日本。「禅」与园林在中国自古就有密切的联系，它们通过相同的精神和不同的表达方式来诠释传统的自然美学。在中国文化五千年的历史长河中，禅学与中国园林一直相互渗透、相互补充、相互影响，两者相辅相成。在中唐之后，明清之前，中国园林一直遵循着禅宗发展的脉络去发展。

Zen (Chan)

Few people [in Chinese] generally write "Chan" [the character the Japanese use for "Zen"]. But I write "Zen" for two reasons: first, there is a taste of "Zen" in the design process; second, Germans think that the Zen school is Japanese because of a translated Japanese book on Zen that made its way to Europe. So Europeans consider Zen to have originated in Japan. Japan's "dry landscaping" [a Japanese style] is very successful, it has a Zen flavor, and is an extension of Zen thinking. But in fact, Zen Buddhism was first introduced to Japan during the Chinese Song Dynasty. "Zen" and gardens have been closely linked in China since ancient times. These two things interpret traditional natural aesthetics in the the same spirit but through different styles of expression. In the 5,000-year-long history of Chinese culture, Chinese gardens and Zen have been continuously permeating, complementing and influencing each other; each assists the other. After the Middle Tang Dynasty and before the Ming and Qing Dynasties, Chinese gardens always developed in the context of the development of Zen.

飒飒西风满院栽，蕊寒香冷蝶难来。

园翁莫把秋荷折，
因与游鱼盖夕阳。

一树春风有两般，
南枝身暖北枝寒。

Some say that Wang Wei's poems concern Zen. It is also said that the paintings of Bada Shanren [17-century painter and calligrapher] have a Zen meaning. How to understand "Zen"? What I have heard is secondhand. But in this photo of a snow scene before the entance I do feel a bit of "Zen" flavor. Zen Buddhism has been in China for more than a thousand years, and later merged with Confucianism and Taoism. It is broad and profound. Don't dare to discuss the Zen or the Tao themselves; one shouldn't even get involved in this question, let alone engage in wild talk. But in garden design, it should be affirmed as a valid pursuit of an artistic idea.

Can we say that Zen is a kind of beauty and of wisdom? Quiet and empty, covering the four seas, eliminating the eight wastelands. Although there is "nothingness" here, is it better to speak in terms of "things"? The aesthetic concept of "less is more" is fully reflected in this idea.

Any matter should be divided into two parts. No matter how profound Zen is, it is impossible to "hang in the void" forever. Once something abstract "lands" in this world it will become targeted by the market. The four characters in "Zen and tea are of the same flavor" are the elegant embellishments in teahouses and restaurants; "House of Zen and Tao" has also become a signboard written in gold characters for a yoga club. It can be seen how far Zen's "folklorization" and "fashionization" have advanced!

Why do I have a soft spot for this picture? It is this atmosphere that we hoped for starting our design. The thinking and process here are different from that of the garden. The long straight lines, the elegant high walls, the dark pool, all set off by a stone or two, boundless imagination and thoughts. And the white snow has summarized and refined the artistic effect of this concept, so I chant for it!

A wise man once wrote in a letter: " 'Under the thatched roof at Yedu, a berth was left for a boat, although no boat will ever come.' This has a 'Zen' 'feeling' to it." Here, although there is no person, no boat, and no paddle, the scene contains a unique figurative beauty. One can use emptiness to better express presence, abstraction to represent the beauty of real things.

In the profundity and depth of the world there is both presence and absence, and implicit expression is better than straightforwardness, so "Zen" is even more the state of things.

行到水穷处，
坐看云起时。

砌下落梅如雪乱，
拂了一身还满。

独坐清谈久亦劳，
碧松燃火暖衾袍。

结束语

中国园林设计与建筑设计一样，都可以表现一种文化，是利用工程技术手段，来艺术地演示一种文化现象。两者相比，就文化属性而言园林更甚。我国造园历史悠久，可被誉为世界艺术之苑中的一株奇葩。就其盛誉，在国外，采用移花接木方式造园，也会使人耳目一新，但模宋仿清，繁雕细琢，不符合本园主题，更非设计者意愿。而以情造园，因地制宜，借其情取其意，根据实地条件才能再现一个理想的境地。

潜园在建造过程中，当地报纸一直跟踪报道，可以说在彼时也算是德国当地的一桩盛事，本书也根据实际情况予以叙述，尽量达到如实回忆。

书中照片经过多次挑选而定，其中有些表面看起来似与园林设计关系不甚直接，纯属艺术照，但很美。因为设计构思就是要在不同场合表现某种层面的美，那些如此细腻和容易被人忽略的角落，是要引起设计者的思考的！

作为结束语，我根据潜园的设计特点归纳为“小”“曲”“借”“度”“意”五个方面，分别概述如下：

小

江南园林也称城市山林，它们都位居城中，占地局限，“小”是共同特点。小中见大就成了中国园林设计的艺术夙求。

方寸之地能容五湖四海，此话虽属文学夸张，但也能促进设计人的艺术构思。如：潜园门前的概括布局、如画的门厅、游廊、主厅、野渡、山谷幽深以及多彩的水面等。诸多景观各具千秋，皆纳入了一亩三分地之中。此乃小中见大的一则实例。水面之上透明如镜的倒影，也适当地扩大了视觉空间。

楹联和匾额是中国园林中独有的装饰物，它不仅点景，好的词句，也会诱人遐想，“远山近水尽收眼底”。俗话说：“人在屋中坐，神入四海游。”——小中见大。

曲

园中游廊曲折蜿蜒，高低起伏，有收、有放妙于变幻。曲廊不仅延长了游赏时间，也会把游人引至更多的美好画面。我们将这一手法誉之为步移景异或曲径通幽。它与小中见大有共通之处，艺术思维不宜机械死板划分。“曲”也是中国园林设计的惯用手法。

借

妙在因借是指借景。借有两种，一是直接的借，如借景。还有一种间接的借，是指设计人借鉴其他艺术，受到启发，衍化而成新的园林景观。例如：墙低更显山高，源自于天坛；破壁残垣，意味着荒远，是受京剧舞台艺术的启发；博物馆的展品和历史件的墙饰，为砖雕似古的景观提供了灵感。潜园设计的立意和构思也是借鉴于《桃花源记》的主题而展开的。“借”是艺术创作的阶梯。“他山之石，可以攻玉。”

度

世间万物皆宜有“度”，本书是指景观处理的恰当程度：入口前的高墙封而不堵，是水洞的玲珑适度所致。游廊中舍去了不必要的装饰，与鸡犬相闻的桃源村舍协调有度。山谷景观，既保持了雄幽兼容的独立个性，又与主厅之间若隐若现，适度呼应。“度”的掌控有如烧菜时加的盐，宜咸淡可口。“度”是衡量一切艺术的标尺。

意

“意”，中国园林设计不仅要视觉美，有时也要意境追求。游廊中留有残缺，耐人寻味，另富含意。野渡景观源自于唐诗，蕴含着诗意。石间鹅卵，寓意曾经的山洪，山外有山，浮想联翩，丰富了境意。山石无言，点立门前以及草棚下的石岸、石屿水中的布局，都是对“禅意”的探寻。这些景观不但要看，也要想，更要悟。

“意”是艺术表象的深层内涵，故而求索之！

CONCLUDING WORDS

Chinese garden design, like architectural design, can be an expression of culture. They are both artistic demonstrations of cultural phenomena using engineering and technological means. Comparatively, gardens are even deeper as cultural representations. China has a very long history of gardening, and it can itself be regarded as a distinctive plant in the garden of world art. As far as its reputation is concerned, it is refreshing to transplant flowers and trees to build gardens overseas. However, merely meticulously imitating the garden styles of the Song or Qing Dynasty did not suit the theme of this garden, let alone align with the designer's desires. One should build a garden based on the local conditions, and further use these conditions to find meaning; only according to the actual local environment can one literally and ideally "reproduce" these conditions.

During the construction of the garden, local newspapers were continually following updates. It can be said that at that time, this was a local historic event there. This book also describes the actual situation, and tries to the extent possible to report what actually happened.

The photos in the book have been chosen after multiple rounds of winnowing down, and some of them seem on the surface to have an indirect relationship with the design of this garden. They are purely artistic photos, but they are beautiful. Because the design idea is to express in different locations a particular level of beauty, those delicate and easily overlooked corners must cause the designer to think!

To conclude, below I summarize by discussing five characteristics of the design of the garden: "smallness", "nonlinearity/crookedness", "borrowing", "degree" and "meaning":

Smallness

The Jiangnan garden is known as an "urban forest". Such forests are located in the city but occupy a limited area. "Smallness" is thus a common feature. Seeing bigness in smallness has become an artistic quest in Chinese garden design.

A square inch [measured in traditional Chinese units] of land can hold five lakes and four seas. Although this is a literary exaggeration, it can also boost the designer's artistic conception. This conception includes such things as the general layout of the front of the garden, the picturesque foyer, the gallery, the main hall, the water crossing, the deep valley, and the colorful water surface. Many of the garden's landscapes have their own unique features, all of which are here included in one [Chinese] acre of three-part land area. This is a real example of seeing the "big" in the small. The mirror-like reflection on the water surface further appropriately enlarges the sense of what can be seen.

Poetic couplets and inscribed plaques are unique decorations in Chinese gardens. Not only do they serve to describe the scenery using beautiful language, but they also induce a sort of reverie. "The faraway mountains and the nearby water together make for a panoramic view," as the saying goes: "A man sits at home, and his mind wanders throughout the four seas." — this is "seeing the big in the small".

Nonlinearity/crookedness

The garden gallery is winding, it rises and falls, and in its variety is sweeping, wonderful. The gallery not only extends the tour, but it also gives more beautiful pictures for visitors to enjoy. We refer to these techniques as a walk through a variety of scenes via winding gallery. It has something in common with seeing the big in the small; artistic thinking should not be rigidly divided. Such nonlinearity is also a custom in Chinese garden design.

Borrowing

"Wonder comes from borrowing" refers to borrowing scenery. There are two types of borrowing. One is direct borrowing, such as borrowing scenery [from outside the garden]. There is also an indirect borrowing, which refers to the designer's being inspired by other arts, with these inspirations evolving to form a new landscape. For example, the lower part of the wall serves to present an image of a tall mountain; this originates from the Temple of Heaven. A broken wall implies distant desolation, and is inspired by Peking opera stage art. The museum's exhibits and historically significant wall decorations provide inspiration for landscapes carved from brick that resemble those of antiquity. The conception of the design of the Qianyuan garden was also developed based on the theme of "Peach Blossom Spring". "Borrowing" is the ladder for creating art; "stones from other hills can learn."

Degree

Everything in the world should have its "degree". This book has referred to the appropriate level of landscape treatment: the high wall in front of the entrance is not blocked, because of the exquisite fit of the water gap. Unnecessary decorations were omitted from the gallery, and it is coordinated at the proper level with the crowded Peach Peach Blossom Spring Cottage. The valley landscape not only maintains a personality that is magnificent, appropriate yet distinctive. It blends, but only indistinctly, with the main pavilion, and echoes at the appropriate level. Control of "degree" is like the salt one adds during cooking — salty, but only to taste. "Degree" is a yardstick for measuring all art.

Meaning

As to "meaning", Chinese garden design must not only achieve the visually beautiful, but also sometimes should implement an artistic conception. The gallery is left incomplete, yet is intriguing and meaningful. The Yedu landscape originates from a Tang poem, and implies that poem's meaning. The cobblestones among the rocks symbolize past mountain torrents. On the hill one can let one's mind roam and imagine other mountains, thus broadening one's perspective. The silent stones on the hill, the stones in front of the entrance gate, the rocks on the bank under the thatched hut, and the arrangement of the rock islets in the water are all explorations of "Zen". These landscapes must not just be seen, but also pondered, and realized.

"Meaning" is the deep inner sense that comes from art's external appearance, so seek it!

《桃花源记》

（东晋）陶渊明

晋太元中，武陵人捕鱼为业。缘溪行，忘路之远近。忽逢桃花林，夹岸数百步，中无杂树，芳草鲜美，落英缤纷。渔人甚异之。复前行，欲穷其林。

林尽水源，便得一山，山有小口，仿佛若有光。便舍船，从口入。初极狭，才通人。复行数十步，豁然开朗。土地平旷，屋舍俨然，有良田美池桑竹之属。阡陌交通，鸡犬相闻。其中往来种作，男女衣着，悉如外人。黄发垂髫，并怡然自乐。

见渔人，乃大惊，问所从来。具答之。便要还家，设酒杀鸡作食。村中闻有此人，咸来问讯。自云先世避秦时乱，率妻子邑人来此绝境，不复出焉，遂与外人间隔。问今是何世，乃不知有汉，无论魏晋。此人一一为具言所闻，皆叹惋。余人各复延至其家，皆出酒食。停数日，辞去。此中人语云："不足为外人道也。"

既出，得其船，便扶向路，处处志之。及郡下，诣太守，说如此。太守即遣人随其往，寻向所志，遂迷，不复得路。

南阳刘子骥，高尚士也，闻之，欣然规往。未果，寻病终，后遂无问津者。

主厅的侧墙上镶嵌着《桃花源记》碑刻，为无臂书法家张文佑的「口书」（用口衔毛笔进行创作）。

张文佑

张文佑四岁因触电失去双手，少随书法家耿春林先生学习书法，以口捉笔，研习历代名家碑帖，寒暑不辍。后又游历名山大川，探访名胜古迹。一九九七年应上海真如寺方丈邀请，耗时八月余，以金粉抄就的小楷长卷《妙法莲花经》，长四百余米，现藏于上海真如寺华东第一佛塔的天宫内。二〇〇〇年，德国鲁尔大学将其书写的《桃花源记》镌成碑文立在潜园内，并邀其赴德进行文化艺术交流。其作品多次被作为礼品赠予外国友人。

Zhang Wenyou

Zhang Wenyou lost his hands at the age of four due to receiving an electric shock. He learned calligraphy under the calligrapher Mr. Geng Chunlin by holding the pen in his mouth and studying inscriptions from famous artists in the past. Later, he visited famous mountains, rivers and other places of interest. In 1997, at the invitation of the abbot there he went to Zhenru Temple in Shanghai. The work he did there took more than eight months. The Xiaokai [an older, more complex style of Chinese characters] scroll he copied in gold-powder paint, *The Magic Lotus*, in the end was more than 400 meters long. In 2000, Ruhr University in Germany inscribed his scroll of "Peach Blossom Spring" into a plaque in the Chinese garden and invited him to Germany for cultural and artistic exchange. His works have been given as gifts to foreign friends many times.

"PEACH BLOSSOM SPRING"

Tao Yuanming
Eastern Jin

In the middle of Taiyuan reign of Jin Dynasty, there was a man from Wuling who was a fisherman by trade, was traveling along the edge of a creek, and forgot how long the route was. At one point he suddenly chanced upon a peach-blossom forest. Keeping by the shore he moved hundreds of steps into it. Inside there were no other kinds of trees, and the fragrant grasses were fresh and beautiful; the fallen leaves were a mix of colors, and the fisherman found it quite unusual. He continued onwards, wanting to find the end of the forest.

At the waterhead at the end of the forest was a mountain. It had a small opening, and it seemed like there was light inside it, so he abandoned his boat and went in. At first the path was extremely narrow, so that only one person could pass. After several dozen more paces, it suddenly opened up. The ground was flat and broad, and on it there were nicely arranged houses. There was fertile land, beautiful pools, mulberry trees and bamboo, roads and lanes for travel, and barking of the chickens and dogs could be heard everywhere. Within they were planting, and the men and women all wore ordinary clothing. The elderly and children both seemed joyful and happy.

They saw the fisherman, and were very surprised, asking where he had come from; he told them everything. They invited him to their homes, brought out wine, slaughtered chickens and prepared a meal. Once it became known in the village that this person was there, everybody came to ask him questions. They began by saying, "Previous generations fled the chaos during the Qin Dynasty, bringing their wives, children and villagers to this place; we have never left. And so we have been separated from outsiders ever since." They asked what era it currently was. It turned out they hadn't even heard of the Han Dynasty, let alone the Wei and the Jin. One by one this person was asked about all he knew, and everyone gasped. The other people all wanted to have him to their homes, and each brought out wine and food. He stayed for several days, then said his goodbyes. As he was leaving, some of the people told him: "It will not be worth your while to tell others."

But thereupon the man departed, then reached his boat, and continued on his travels, leaving markers all along the route. Upon reaching the county seat, he went to see the governor, and recounted his story. The governor then sent people to follow his trail, searching for the signs he had left; but they were bewitched and could not find the road.

Liu Ziji of Nanyang was a noble scholar who heard this story, and enthusiastically set out to find this place, but was unsuccessful, dying of illness during his search. After that, no one ever asked again.

明代仇英绘《桃花源图》（局部）
资料来源：美国波士顿美术馆藏

大事记

1.1990 年 8 月 7 日，《西德意志汇报》下属《波鸿导报》第 181 期第 1 版刊登照片，展示鲁尔大学植物园中正在修建的潜园工地，称该园将于秋季完工；第 4 版以《鬼怪也要绕路：中国园林开工 有宝塔和柳树》对此作了详细报道。

2.1990 年，《鲁尔新闻报》也刊登照片，以《中国园林初具规模》为题作了报道，并称该园将于 10 月底开放。

3.1990 年 10 月 5 日，《西德意志汇报》下属《波鸿导报》第 232 期第 6 版以《中国园林中充满符号：丰饶和财富 大学景点 11 月完工 山洞和石穴》为题报道了潜园的建设情况。

4.1990 年 11 月 29 日，潜园举行开园仪式，鲁尔大学校长沃尔夫冈 · 马斯贝格、波鸿市市监迪特尔 · 邦格特、波鸿市市长海因茨 · 艾克尔贝克，以及来自同济大学的代表出席。

5.1990 年 11 月 30 日，《西德意志汇报》下属《波鸿导报》第 280 期第 3 版以《艺术作品的一切道路源自日常生活：鲁尔大学中国园林开园》为题对此做了报道。

6.1990 年 12 月 17 日，波鸿鲁尔大学校报 *RUB-Aktuell* 第 1 版刊登照片，并报道了潜园开放的消息。

7.1991 年上半年，波鸿鲁尔大学植物园的技术负责人贝恩德 · 基希纳以《波鸿鲁尔大学的中国园林》为题介绍了造园项目的缘起（1990 年 5 月至 11 月鲁尔大学与同济大学合作项目的成果）、建造过程和艺术特色。

8.2000 年，波鸿市政府主办的地方志刊物 *Bochumer Zeitpunkte* 第 7 辑转载了贝恩德·基希纳的文章《波鸿鲁尔大学的中国园林》，并配有建筑师马丁 · 拜尔曼的一段导读。

9.2000 年，在波鸿鲁尔大学成立“中国园林协会”，负责潜园的维护修缮工作。

10.2001 年 10 月，潜园重新开放。

11.2013 年 7 月 27 日，《西德意志汇报》以《中国园林需要修缮》为题介绍了鲁尔大学植物园中的中国园需要修缮。

12.2014 年 5 月 20 日，《西德意志汇报》以《中国园林需要中国工艺》为题介绍了鲁尔大学植物园的中国园修缮工程。

13.2014 年 7 月 15 日，《鲁尔新闻报》以《中国园林已修缮》为题介绍了中国工匠们小心翼翼地修复了鲁尔大学里的中国园林。

14.2019 年 5 月至 8 月，中国园林协会于每月的第 3 个星期六提供每次时长 1 小时的免费公众导览服务。

Documentation of Major Events

1. On August 7, 1990, the first page of the 181st issue of the *Bochumer Anzeiger*, a regional publication of the *Westdeutsche Allgemeine Zeitung* (*WAZ*), published a photo showing the construction site of the Qianyuan Garden in the Botanical Garden of Ruhr University, and said the garden would be completed in the fall. A detailed report was included on the fourth page, under the title "Bad Spirits Must Take a Detour: Construction Work Has Started on the Chinese Garden's Pagodas and Willows".
2. In 1990, *Ruhr Nachrichten* also included a photo in a report titled "The Chinese Garden is Taking Shape", and said that the garden would open at the end of October.
3. On October 5, 1990, the 6th page of issue 232 of *Bochumer Anzeiger*, a regional publication of the *Westdeutsche Allgemeine Zeitung* (*WAZ*), included the article "Chinese Gardens Full of Symbols of Fertility and Wealth; Work on Niches and Grottoes Will be Completed in November".
4. On November 29, 1990, the opening ceremony for the garden was held. The president of Ruhr University Wolfgang Maßberg, Bochum City Supervisor Dieter Bongert, Bochum Mayor Heinz Eikelbeck and representatives from Tongji University attended.
5. On November 30th,1990, the 3rd page of the 280th issue of the *Bochumer Anzeiger*, a regional publication of the *Westdeutsche Allgemeine Zeitung* (*WAZ*), included the article "For Art, All Paths Come from Everyday Life: The Chinese Garden at Ruhr University Has Opened".
6. On December 17, 1990, the first page of the Ruhr University newspaper, *RUB-Aktuell*, published photos and reported on the opening of the garden.
7. In the first half of 1991, Bernd Kirchner, the technical director of the Botanical Garden of Ruhr University, Bochum, in an article titled "The Chinese Garden of Ruhr University, Bochum", introduced the origin (as the fruit of a cooperative project between Ruhr University and Tongji University from May-November, 1990), construction process and artistic features of the garden.
8. In 2000, the seventh issue of the Bochum municipal government's *Bochumer Zeitpunkte* republished Bernd Kirchner's article "The Chinese Garden of Ruhr University, Bochum", along with a guide by architect Martin Beilmann.
9. In 2000, the "Chinese Garden Association" was established at Ruhr University in Bochum to take over the garden's maintenance and repair.
10. In October 2001 the park reopened.
11. On July 27, 2013, the *Westdeutsche Allgemeine Zeitung* (*WAZ*) reported in an article titled "Chinese Garden Needs Renovation" on the need to repair the Chinese garden in the Botanical Garden of Ruhr University.
12. On May 20, 2014, the *Westdeutsche Allgemeine Zeitung* (*WAZ*) in an article titled "Chinese Garden Needs Chinese Craftwork" introduced the renovation project for the Chinese garden at Ruhr University.
13. On July 15, 2014, *Ruhr Nachrichten* in an article titled "Chinese Garden Renovated" discussed the careful repair by Chinese craftsmen of the garden at Ruhr University.
14. From May to August 2019, the Chinese Garden Association provided one-hour free guided tours for the public on the 3rd Saturday of each month.

图书在版编目（CIP）数据

画谈潜园 : 中国园林在德国 : 汉英对照 / 张振山著 ; (美) 欧思博 (Evan Osborne) 译 . -- 上海 : 同济大学出版社 , 2020.6

ISBN 978-7-5608-8931-3

Ⅰ . ①画… Ⅱ . ①张… ②欧… Ⅲ . ①园林设计－中国－图集 Ⅳ . ① TU986.2-64

中国版本图书馆 CIP 数据核字 (2020) 第 095248 号

画谈潜园：中国园林在德国

张振山　著　[美] 欧思博　译

策划编辑　江　岱
责任编辑　张　微
责任校对　徐春莲
书籍设计　张　微
出版发行　同济大学出版社　www.tongjipress.com.cn
（地址 上海市四平路 1239 号 邮编 200092 电话 021-65985622）
经　　销　全国各地新华书店
印　　刷　上海雅昌艺术印刷有限公司
开　　本　889 mm× 1194mm 1/16
印　　张　11
字　　数　352 000
版　　次　2020 年 6 月 第 1 版　　2020 年 6 月 第 1 次印刷
书　　号　ISBN 978-7-5608-8931-3
定　　价　138.00 元

图片提供
波鸿市中国园林协会

亦石亦画
Of Rocks and Paintings

环保之歌
Song of Environmental Protection

风月亭记
An Account of the Fengyue Pavilion

石水之间
Between Rocks and Water

山不在高
A Mountain Need Not Be Tall

文化乐园
A Cultural Paradise

中国茶室
A Chinese Tea House

外墙景观
Exterior Landscapes

瓦的故事
The Story of the Tiles

妙在因借
Borrowed Wonder

水绘春秋
Water Paintings, Spring and Autumn

姹紫嫣红
Beautiful Flowers of All Sorts

银装素裹
Silver Adornments, Plainly Wrapped

问道于禅
Asking about the Tao and Zen (Chan)